NUCLEAR ENGINEERING DATA BASES, STANDARDS, AND NUMERICAL ANALYSIS

NUCLEAR ENGINEERING DATA BASES, STANDARDS, AND NUMERICAL ANALYSIS

Jacek Jedruch

VNR VAN NOSTRAND REINHOLD COMPANY
New York

Library of Congress Catalog Card Number: 84-20970
ISBN: 0-442-24574-2

Manufactured in the United States of America

Published by Van Nostrand Reinhold Company Inc.
135 West 50th Street
New York, New York 10020

Van Nostrand Reinhold Company Limited
Molly Millars Lane
Wokingham, Berkshire RG11 2PY, England

Van Nostrand Reinhold
480 Latrobe Street
Melbourne, Victoria 3000, Australia

Macmillan of Canada
Division of Gage Publishing Limited
164 Commander Boulevard
Agincourt, Ontario M1S 3C7, Canada

15 14 13 12 11 10 9 8 7 6 5 4 3 2 1

Library of Congress Cataloging in Publication Data

Jedruch, Jacek.
Nuclear engineering data bases, standards, and numerical analysis

Bibliography: p.
Includes index.
1. Nuclear engineering—Information services.
2. Information storage and retrieval systems—
Nuclear engineering. 3. Data libraries. I. Title.
TK9153.J44 1985 025′.0662148 84-20970
ISBN 0-442-24574-2

To my wife Eva

PREFACE

During the last quarter century a large nuclear industry has grown to serve the needs of the United States in producing energy. Its establishments are easily seen: uranium mines and processing plants, national laboratories, design and manufacturing facilities of the four major reactor suppliers, power plants generating electricity. Less known, but equally important, is the store of information, data, computing methods, regulations and standards which grew alongside this industry. The present book focuses on the sources and repositories of this "software of nuclear engineering," and the methods of information transfer between them. It is intended as a guide for practicing nuclear engineers, personnel of public utilities, state and federal regulatory bodies, and students of nuclear engineering.

The existence and easy availability of many excellent books on reactor theory tend to obscure the scarcity of texts on the content, location and use of data bases, the choice of available numerical methods, standards and regulations needed to translate reactor theory into correct designs or operating practice. To remedy this deficiency, the present work emphasizes the implementation of methods of reactor engineering, rather than the reactor theory itself. It outlines the most significant parts of this body of information, and by going into its description in sufficient detail, makes it easier for a prospective user to find his way.

The data base and methods of nuclear engineering possess many unique features and specific strengths. They embody some of the most advanced techniques of applied mathematics, numerical analysis, mechanized data evaluation and handling, and rely on the cooperative organization of data base management. The principal collections of data, codes, standards and regulations deserve to be recognized and identified for their usefulness and applicability.

The first major block of the "software of nuclear engineering" is its data base. It is defined broadly here as the pertinent numerical and procedural information derived from various related experimental sources, and supported by a large body of technical literature. In the course of reactor design, the material contained in the data base becomes the numerical or factual input to machine calculations. In turn, the computer codes developed during the past three decades to handle calculations embody an essential part of the methods of nuclear engineering, and constitute the second major part of the "software" mentioned above. These numerical methods represent the encoded computational techniques for converting the information contained in the data base into practical design. Finally, the third major part is made up of all engineering standards, federal regulations and guides that were developed over the years to assure maintenance of safety and high quality of products and services offered by the nuclear industry. A large cooperative effort, involving thousands of leading specialists in their fields, was needed to establish this part of nuclear technology.

It is worthwhile to pause here and identify the sources and the means by which the flow of information is accomplished within the field. The experimental facilities and reactors may be viewed in broad terms as the source of the data base. Experimental programs, carried out to verify the reactor theory, produce voluminous accumulation of data, documented in thousands of technical reports. Once the reactors are built and put into operation, information about their engineering, safety and economic performance starts to accumulate. The institutions and facilities that store this information, update and evaluate it, and ultimately disseminate it to users, become a vital part of the system. Engineering societies, regulatory bodies, and commercial publishers interpret, analyze and give formal expression to the data.

One aspect of the operation of many present day data bases deserves particular attention, namely their gradual transformation from static accumulations of information to a dynamic, interactive form. This latter transformation is a direct outcome of the ongoing revolution in microcircuitry, electronics and computer science. With the advent of highly reliable transmission systems using public telephone lines, more powerful computing machinery and storage devices, reasonably priced and reliable display terminals and small computers, it has become practicable for many heretofore static repositories of data to offer on-line interactive data base access.

Related developments taking place at present include the efforts to establish an efficient national data base management system. Direct access to the various data collections by telephone link, with minimal intervention by the center's staff, is frequently possible from terminal points at designated authorized installations throughout the country. There is a dramatic expansion of capabilities in this area, and their significance is only now getting to be appreciated in the extension of the range and accessibility of information available to the designer and operator of nuclear facilities.

The resources described in the present work are available to qualified users in the public domain, generally at the cost of reproduction and handling, or for a user fee. The much more expensive proprietary products, available for purchase at prices established through specific contracts, are not discussed in any detail to avoid infringement on proprietary rights. The availability of a product or a service in the public domain generally indicates that it was developed through a federally sponsored project. In fact, many of the data bases or computer programs, both in and outside the public domain, represent major feats of analytical and programming ability, as well as many man-years of effort.

As an illustration of the advances in automated information handling, it may be mentioned here that the text of this book was written in its entirety with the aid of automated systems. Much material for this work was obtained through on-line searches of several data banks. The text was then keyed into the IBM Personal Computer, edited by means of a word processor. Spelling and grammar were checked with the aid of suitable programs, and the final text was made available to the publisher on magnetic disks for phototypesetting.

Pittsburgh, PA

JACEK JEDRUCH

ACKNOWLEDGMENTS

I wish to thank the following persons for their willingness to review, comment, and in some cases supplement the material contained in the sections on the national data bases. In particular, appreciation is due to Robert W. Roussin of the Radiation Shielding Information Center at Oak Ridge National Laboratory; Robert J. Howerton, Physical Data Group, Lawrence Livermore National Laboratory; Margaret Butler, National Energy Software Center, Argonne National Laboratory; and Sol Pearlstein, National Nuclear Data Center, Brookhaven National Laboratory. Robert M. Westfall of ORNL has reviewed the section on code benchmarking activities. The assistance and cooperation of Paul Allmendinger, June Ling and C. J. Sullivan of the American Society of Mechanical Engineers in reviewing their activities in the area of codes and standards is gratefully acknowledged. Thanks also to John J. Taylor and Arthur F. Billy of the Electric Power Research Institute for supplementing the information concerning the data bases and the activities of EPRI.

Charles L. Nalezny of Idaho National Engineering Laboratory has made available to me the latest materials on LOFT experiments, and Herbert Hecker of Bettis Atomic Power Laboratory pointed out additional materials in the public domain on the Light Water Breeder Reactor. The section dealing with CINDAS property data center at Purdue University was reviewed by C. Y. Ho. I appreciate their help.

And last, but not least, my sincere thanks to Alberta Gordon and Connie MacDonald of Van Nostrand Reinhold Company for their untiring efforts in editing the manuscript and managing the book production.

ACKNOWLEDGMENTS

ACRONYMS APPEARING IN THE TEXT

ACRR	Annular Core Research Reactor (SNL)
ACRS	Advisory Committee on Reactor Safety (NRC)
ACM	Association for Computing Machinery
ALARA	As Low As Reasonably Achievable radiation dose (generic term)
ANL	Argonne National Laboratory (Argonne, IL)
ANP	Aircraft Nuclear Propusion (generic term)
ANS	American Nuclear Society
ASCII	American Standard Code for Information Interchange
ANSI	American National Standards Institute
ATWS	Anticipated Transient Without Scram (generic term)
BAPL	Bettis Atomic Power Laboratory (Pittsburgh, PA)
BNL	Brookhaven National Laboratory (Long Island, NY)
BOP	Balance of Plant, outside NSSS (generic term)
BWR	Boiling Water Reactor (generic term)
CBPC	Computational Benchmark Problem Committee (ANS)
CSD	Computer Sciences Division, Union Carbide (ORNL)
CSEWG	Cross Section Evaluation Working Group (BNL)
CINDAS	Center for Information, Numerical Data Analysis and Synthesis (Purdue University, IN)
CRBR	Clinch River Breeder Reactor (Oak Ridge, TN)
CFR	Code of Federal Regulations
DOD	Department of Defense
DOE	Department of Energy
EBR	Experimental Breeder Reactor (INEL)
ECC	Emergency Core Cooling (generic term)
EG&G	Edgerton, Germeshousen & Grier, Inc. (Idaho Falls, ID)
ENDF	Evaluated Nuclear Data File (BNL)
ENSDF	Evaluated Nuclear Structure Data File (ORNL)
EPRI	Electric Power Research Institute (Palo Alto, CA)
ERDA	Energy Research and Development Administration
ETEC	Energy Technology Engineering Center (formerly LMEC)
FFTF	Fast Flux Test Facility (HEDL)
FLECHT	Full Length Emergency Core Heat Transfer (Pittsburgh, PA)
FPDB	Fuel Performance Data Base (EPRI)

FR	Federal Register
GA	General Atomics Corporation (San Diego, CA)
GCFR	Gas Cooled Fast Reactor (generic term)
GE	General Electric Corporation
HCDA	Hypothetical Core Disruptive Accident (generic term)
HEDL	Hanford Engineering Development Laboratory (Richland, WA)
HFIR	High Flux Isotope Reactor (ORNL)
HTGR	High Temperature Graphite Reactor (generic term)
ICONS	Information Center On Nuclear Standards (ANS)
INEL	Idaho National Engineering Laboratory (Idaho Falls, ID)
KAPL	Knolls Atomic Power Laboratory (Schenectady, NY)
LANL	Los Alamos National Laboratory (formerly LASL)
LASL	Los Alamos Scientific Laboratory (Los Alamos, NM)
LER	Licensee Event Report (generic term)
LLNL	Lawrence Livermore National Laboratory (formerly LLL)
LLW	Low Level Waste (generic term)
LMEC	Liquid Metal Engineering Center (Canoga Park, CA)
LMFBR	Liquid Metal Fast Breeder Reactor (generic term)
LOCA	Loss of Coolant Accident (generic term)
LOFT	Loss of Fluid Test (INEL)
LWR	Light Water Reactor (either PWR or BWR), (generic term)
MAGI	Mathematical Applications Group (Elmsford, NY)
MFECC	Magnetic Fusion Energy Computer Center (LLNL)
NESC	National Energy Software Center (ANL)
NERVA	Nuclear Engine for Rocket Vehicular Application (NRDS)
NNDC	National Nuclear Data Center (BNL)
NPRDS	Nuclear Plant Reliability Data System (INPO)
NRC	Nuclear Regulatory Commission
NRDS	Nuclear Rocket Development Station (Nevada)
NSA	Nuclear Science Abstracts
NSAC	Nuclear Safety Analysis Center (EPRI)
NSMB	Nuclear Standards Management Board (ANSI)
NSSS	Nuclear Steam Supply System (generic term)
NTIS	National Technical Information Service (Springfield, VA)
OECD	Organization for Economic Cooperation and Development
ORELA	Oak Ridge Electron Linear Accelerator (ORNL)
ORNL	Oak Ridge National Laboratory (Oak Ridge, TN)
PRA	Probabilistic Risk Assessment (analytic method)
PTS	Pressurized Thermal Shock (generic term)
PWR	Pressurized Water Reactor (generic term)
RECON	Remote Console system (TIC)

SACRD	Safety Analysis Computerized Reactor Data (ORNL)
SEASET	Separate-Effects and System-Effects Test
SEFOR	South East Fast Oxide Reactor
SLSF	Sodium Loop Safety Facility (INEL)
SNAP	Space Nuclear Auxiliary Power (reactor in earth orbit)
SNL	Sandia National Laboratory (Albuquerque, NM)
SRO	Senior Reactor Operator (generic title)
SRP	Savannah River Plant (SRL)
SRL	Savannah River Laboratory (Aiken, SC)
THTF	Thermal Hydraulic Test Facility
TIC	Technical Information Center (ORNL)
TMI	Three Mile Island (PA)
TPRC	Thermophysical Properties Research Center (Purdue U.)
TREAT	Transient Reactor Test (ANL)
TSF	Tower Shielding Facility (ORNL)
UW	University of Wisconsin
WNES	Westinghouse Nuclear Energy Systems (Pittsburgh, PA)
ZPPR	Zero Power Plutonium Reactor (INEL)
ZPR	Zero Power Reactor (INEL)

CONTENTS

LIST OF ILLUSTRATIONS

NUCLEAR ENGINEERING DATA BASES, STANDARDS, AND NUMERICAL ANALYSIS

1. EXPERIMENTS OF NUCLEAR REACTOR ENGINEERING

1.1 SOURCES OF NUCLEAR AND SHIELDING DATA

The design of present-day nuclear reactors rests on the know-how of the currently practicing nuclear engineers and scientists, as well as the large store of information derived from past activities in the field and subsequently accumulated in the data bases of the industry. Before we delve into a description of the contents of the various specialized data bases of this industry, a brief survey of the experimental sources of the data contained in them is in order. A detailed description of the experimental techniques and the methods of collecting and analyzing data is outside the framework of the present study. Instead, a condensed guide to the experimental sources is provided in this chapter. In taking this approach it is recognized that the experimental techniques used to gather data for reactor design constitute a large and specialized field, partly outside the field of nuclear engineering proper.

The bulk of experimental data currently generated falls into two major categories: nuclear data and reactor performance and safety data. The first of these categories includes neutron, gamma and charged particle cross-section measurements plus lattice and solution criticality experiments, as well as shielding experiments. Sustained experimental programs in this area are being carried out at a steady pace, with the bulk of available funding going into high energy physics and investigations into the structure of matter. The latter area is outside the scope of the present study.

Experimental data on reactor safety and performance come from several major experimental series planned and executed in past years, and some continuing currently. These experimental programs are being carried out in support of the development of pressurized water reactors, the fast breeder reactor program and the fusion program.

Starting with the experiments or observations of operating systems, the time sequence of events leading from the initial recording of the data to the general availability of the information derived from them includes first, as a rule, the publication of a laboratory report. A great deal of the information that eventually finds its way into the various data bases first sees the light of day in topical reports. These are often, but not always, followed by the presentation of results of a given study at a society meeting, or publication of a paper in a technical journal. When enough material on a given topic has accumulated, a monograph dealing with the matter may be published, or the results of many studies may be discussed in one of the annual reviews.

The importance of topical reports has been recognized by the federal government, by providing authorization and financial resources for the establishment of the National Technical Information Service (NTIS) center in Springfield, VA. NTIS serves as a repository of over a million of these reports. Most of them are unclassified, and are available for a fee to practicing engineers or other members of the public. The function of topical reports, as a timely channel for meeting the immediate needs of the practicing nuclear engineer, was also recognized by the American Nuclear Society. In 1975 it initiated the publication in the *Nuclear Science and Engineering* of a list of titles and brief descriptions of new topical reports reaching NTIS each month from the various national and industrial laboratories supporting the nuclear industry.

Cross-Section Experiments

The measurements of cross sections for interaction of neutrons and photons with matter constitute the largest single block of data in the data base of nuclear engineering, and the legacy of reports and know how generated constitutes a substantive part of the present-day understanding of the structure of matter. Virtually all analytical methods of reactor design use as input some of the results of these experiments. The experiments in question took anywhere from months to years to carry out, and involved the participation of sizable staffs of experimental and theoretical reactor physicists. They were carried out by several experimental methods, the most significant being the neutron chopper experiments, the accelerator experiments, and the underground fission bomb experiments. Although a good part of these experiments took place at national laboratories, industrial and university research laboratories contributed as well.

The work in this field is still continuing for materials and isotopes heretofore not investigated. As further improvements are made in the accuracy of experimental methods, cross sections first measured in the early days are being remeasured, or the range of energy covered is being extended.

In the very early days of the development of nuclear engineering, the measurement of material cross sections for interaction with neutrons was much hampered by a lack of intense neutron sources. With the arrival of chain-reacting assemblies, or piles, as they were called, copious neutron sources became available for measurement of cross sections. These early neutron sources, as well as some of the present-day sources, such as that obtained from the High Flux Beam Reactor of Brookhaven, are characterized by a wide distribution of neutron energies. A useful description of the early measurement methods was given by Donald Hughes in his *Pile Neutron Research.*[(1)] A good introductory description of the more up-to-date methods can be found in Karl Wirtz's book *Neutron Physics,*[(2)] and current state-of-the-art descriptions of experimental methods will be found in the proceedings of symposia held from time to time by experimentalists active in this field.[(3)]

In consequence of the broad distribution of neutron energies produced by fission sources, the better part of experimental effort in measuring cross sections with these sources was directed at the selection of neutron energies, commonly done by the neutron pulse chopping technique, or simply "chopper."

In the chopper technique, multienergetic neutrons originating from the interior of a nuclear reactor, or a weapons-type neutron burst, are collimated by a suitable arrangement of shields and absorbers, to pass in a narrow beam in the path of a rotating absorber. The absorber either obstructs or reveals for the neutron beam a long, evacuated flight passage. At the end of the passage samples of the material whose cross sections are to be measured are placed in front of a neutron-detecting apparatus. The attenuation of the incident neutron beam by absorption in the material sample measured, plus knowledge of the length of the flight path and the frequency of rotation of the chopper rotor, gives the requisite information on the energy of the interacting neutrons, and the cross sections measured.

An alternate method of cross-section measurement uses neutrons produced from reactions between charged fundamental particles and matter. Particle accelerators used in this method permit a selection of a much narrower range of neutron energies through the proper choice of target and energy of the incident charged particles. Neutrons generated by a reaction in the target are collimated in the conventional manner, and are used in the measurement of cross sections as usual. In this field of measurement the Oak Ridge Electron Linear Accelerator (ORELA) plays a leading role by providing an intense neutron source.

An important addition to the choices of measuring neutron cross sections by the time-of-flight method is the use of underground explosions[(4),(5)] of nuclear devices to provide very strong sources of neutrons. Part of the neutron pulse derived from the explosion is passed through a collimating vertical tube to the surface of the ground, deflected by a suitable scatterer, and directed at a rotating wheel carrying samples of isotopic species to be measured. This method is particularly useful for measuring the fission cross sections of isotopes available only in small quantities.

An additional method of deriving values of cross sections for fissile and fertile isotopes of heavy metals is to irradiate samples of interest to high burnups in fission reactors of well known neutron spectrum. Isotopic analysis of the discharge material, followed by calculation, allows one to deduce from the results effective values of the actinide cross sections. This approach has led E. J. Hennely to derive multigroup cross section sets (37 group and 84 group) for heavy metal chains,[(3)] based on the evaluated nuclear data, reactor isotope production experiments, multigroup production codes, and cross-section systematics theory.

The data for the derivation of transmutation constants of the elements in the actinide group initially became available in 1960 from the isotopic analysis of the discharged first core loading of the Yankee-Rowe Reactor. In the subsequent analysis, constants relevant to the conversion of lower into higher members of the actinide chains were extracted from the isotopic data.[(6),(7)] This was followed by studies performed at Oak Ridge, using the High Flux Isotope Reactor[(8)] to carry out required irradiations, and at the Savannah River Laboratory through fuel sample irradiations in the Savannah River Plant (SRP) production reactor.[(9)] Data obtained in the above experiments are of special value in confirming calculations connected with the operation of reactor fuel cycles, and in studies supporting the development of methodologies of reactor safeguards. In the last application, accurate predictions of the composition of discharged fuel are necessary to prevent diversion of special nuclear materials to unauthorized uses.[(10)]

REFERENCES

1. D. J. Hughes, *Pile Neutron Research*, Addison-Wesley Publishing Company, Cambridge, MA (1953).
2. K. H. Beckurts and K. Wirtz, *Neutron Physics*, Springer-Verlag, New York (1964).
3. E. J. Hennely, "The Heavy Actinide Cross Section Story," *Proc*. Third Conf. on Neutron Cross Sections and Technology, Mar. 15–17, 1971, Knoxville TN, Conf-710301, Vol. 2, p. 494 (1971).
4. Gary H. Higgins, "Summary of Nuclear Explosion Data for Underground Engineering Applications," UCRL-50853, Lawrence Radiation Lab., Livermore, CA (1970).
5. M. S. Moore and G. A. Keyworth, "Analysis of the Fission and Capture Cross Sections of the Curium Isotopes," LA-DC-12230, Los Alamos Scientific Lab., Los Alamos, NM (1970).
6. J. Jedruch, "Statistical Analysis of Alpha-25 Obtained from the Destructive Analysis of

Spent Yankee Fuel," WCAP-6057, Westinghouse Electric Corporation, Pittsburgh, PA (1963).
7. J. Jedruch and R. J. Nodvik, "Determination of Core Burnup, Isotopic Inventory and Conversion Ratio from Selective Sampling of Spent Fuel, *Nuclear Applications*, 3, 507–518 (1967).
8. L. J. King, J. E. Bigelow, and E. D. Collins, "Transuranium Processing Plant Semiannual Report," ORNL-5034, Oak Ridge National Lab., Oak Ridge, TN (1975).
9. R. W. Benjamin, F. J. McCrosson, T. C. Gorrell, and V. D. Vandervelde, "A Consistent Set of Heavy Actinide Multigroup Cross Sections," DP-1394, Savannah River Lab., Aiken, SC (1975).
10. H. Ottmar and C. Weitkamp, "Nuclear Data for Safeguards. Can Better Data Improve Present Techniques?" Symp. on Practical Applications of Research and Development in the Field of Safeguards, Rome (Mar. 1974).

Lattice Experiments

Associated with the nuclear data base is a substantial body of information produced in reactor physics experiments, carried out to verify the results of calculational predictions. These experiments range from early subcritical assemblies, used to verify lattice constants, through critical assemblies set up to support the demonstration reactor program, all the way to the most recent mockups of entire reactor cores for breeding of fissile materials and for investigating the fusion environment.

Summaries of data obtained in the physics experiments were included in such reports as the very popular Argonne "Reactor Physics Constants," ANL-5800.[1] This report was initially published in 1958 in soft covers. It enjoyed such popularity among reactor designers that it was enlarged, revised and reissued as a large volume in hard covers in 1963. The data from lattice experiments were also extracted and published in a three volume British report, "Handbook of Experimental Criticality Data."[2] From time to time, various bibliographies of lattice experiments supporting reactor physics have been published.[3]–[7] Brian L. Koponen recently produced a compilation of criticality experiments.[8]

Of the presently operating critical facilities, perhaps the most active is the one at Los Alamos National Laboratory.[9] Its criticals have been in operation for 37 years, providing information in the areas of integral cross sections and criticality safety. Several criticality machines are available at that facility. They are:

- PLANET machine, used to demonstrate stacking of critical assemblies by hand, using alternating lucite and heavy metal foils. The effect of source and detector configurations on the measurement of reciprocal multiplication can be studied.
- SHEBA, a uranyl-fluoride solution assembly, used to demonstrate geo-

metric effects on multiplication. It can be used in the runaway mode in demonstrations of accident conditions and reaction self-quenching effects through thermal expansion.

- HONEYCOMB, a machine used for studying UO_2 beryllium reflected assemblies, including replacement measurements, control rod calibration, and the effect of reflector thickness.
- FLATTOP machine, a metallic plutonium core, reflected with natural uranium; used to demonstrate control rod calibration and interactions, as well as replacement measurements.

Also available at the Los Alamos Laboratory, is the GODIVA-IV bare metal critical assembly, which can be operated in the super-prompt critical burst mode.

A large block of fast reactor physics experiments was done with the split table Zero Power Reactor (ZPR). In this experimental arrangement, each of the two matching parts of a metal honeycomb contains one half of the core mockup in the form of drawers filled with fuel, or structure-coolant materials. When brought together, the complete assembly becomes critical. The ZPR facility was constructed in 1955 at the Idaho National Reactor Testing Station to accommodate fast reactor critical experiments of the Argonne National Laboratory. Under sequential names ZPR-1 through ZPR-9, it hosted critical experiments for the EBR-II reactor, French Fast Reactor RAPSODIE with U–Pu–Mo fuel elements containing up to 430 kg Pu, and the SEFOR criticals. Large LMFBR criticals, with 3100-liter core and 1120 kg critical mass, and the GCFR benchmark assemblies were also tested.(10)

When the workload at the ZPR facility exceeded its capacity in the late 1960s, a companion facility, the Zero Power Plutonium Reactor (ZPPR), was erected to house critical mockups of reactor-like critical systems, while supporting measurements on closely related core compositions were relegated to the ZPR split table machine. Consecutive critical assemblies named ZPPR-1 through ZPPR-11 contained experiments for the 350 MW(e) Clinch River Breeder Reactor, an LMFBR scale up to 700 MW(e) size, as well as expanded cores of 4600-liter volume and 2000 kg of fissile Pu content, corresponding to a 900 MW(e) LMFBR.(11)

Recent experimental work with lattices is concerned mostly with the prevention of criticality of densified spent fuel storage racks. This is a practical problem arising out of delay in developing a coherent national policy for storage and reprocessing of spent reactor fuel and radioactive waste. Without an adequate number of spent fuel repositories for away-from-reactor storage, utility companies operating reactors have been forced to pack more spent fuel into existing storage facilities by decreasing the separation between the stored fuel assemblies. To accomplish this operation without causing inadvertent criticality, analytical and experimental confirmation of the proposed fuel configurations is required. Lattice experiments needed for this purpose have been carried out [(12)–(14)] at Babcock and Wilcox, Oak Ridge National Laboratory and Pacific Northwest Laboratory.

Compared to the early decades in the development of the nuclear industry, there is a noticeable decrease in the number of lattice and critical experiments undertaken today. This reflects increased confidence in analytical methods, and justifies an early saying: "the purpose of critical experiments is to have no critical experiments." The early intense activity in this field has produced several monographs dealing with experimental techniques.[(15)–(17)]

REFERENCES

1. L. J. Templin, ed., "Reactor Physics Constants," ANL-5800, Argonne National Lab. (July 1963).
2. "Handbook of Experimental Criticality Data," United Kingdom Atomic Energy Authority, AHSB (S) Handbook 5, Parts I–III (1967–1968).
3. R. L. Scott, "Criticality Studies (A Literature Search)," TID-3533, U.S. Atomic Energy Commission (Aug. 1959).
4. W. E. Bost, "Fast Neutron Pulse Reactors and Fast Neutron Critical Assemblies (A Literature Search)," TID-3571, U.S. Atomic Energy Commission (Apr. 1962).
5. Jean Furnish, "The Data of Nuclear Reactor Physics. A Bibliography," LA-3740, Los Alamos Scientific Lab. (Apr. 1967).
6. Jean Furnish, "Data of Nuclear Reactor Physics, 1967–1968. A Bibliography," LA-4225, Los Alamos Scientific Lab. (Sept. 1969).
7. H. C. Paxton, "Experimental Criticality Specifications, An Annotated Bibliography Through 1976," LA-UR-77-684, Los Alamos Scientific Lab. (1977).
8. Brian L. Koponen, Thomas P. Wilcox, and Victor E. Hampel, "Nuclear Criticality Experiments from 1943 to 1978. An Annotated Bibliography: Vol. 1, Main Listing; Vol. 2, Lookup Tables; Vol. 3, Subject Index," UCRL-52769, Lawrence Livermore National Lab. (1979)
9. R. A. Malenfant, "Los Alamos Critical Assemblies Facility," LA-8762-MS, Los Alamos National Lab. (June 1981).
10. C. E. Till, and W. G. Davey, "The Demonstration Reactor Benchmark Critical Assemblies Program," Trans. ANS, 13, 293–294 (June 1970).
11. L. G. LeSage, M. L. Lineberry, H. F. McFarlane, S. G. Carpenter, and P. I. Amundson, "A Benchmark Critical Experiment Program for 600- to 700-MW(e) LMFBRs," *Trans. ANS*, 33, 834–836 (Nov. 1979).
12. G. S. Hoovler, M. N. Baldwin, E. L. Maceda, and F. G. Welfare, "Critical Experiments Supporting Underwater Storage of Tightly Packed Configurations of Spent Fuel Pins," BAW-1645-4, Babcock & Wilcox Co. (Nov. 1981).
13. S. R. Bierman, B. M. Durst, and E. D. Clayton, "Critical Separation Between Subcritical Clusters of Low-Enriched UO_2 Rods in Water with Fixed Neutron Poisons," *Nucl. Technol.*, 42, 237 (1979).

14. J. T. Thomas, ed., "Nuclear Safety Guide," TID-7016, Rev. 2, NUREG/CR-0095, ORNL/NUREG/CSD-6, Oak Ridge National Lab. (1978).
15. Joseph A. Thie, *Heavy Water Exponential Experiments Using ThO_2 and UO_2*, Pergamon Press, Elmsford, NY (1961).
16. Loren C. Schmid, *Critical Assemblies and Reactor Research*, John Wiley & Sons, New York (1971).
17. A. Edward Profio, *Experimental Reactor Physics*, John Wiley & Sons, New York (1975).

Tank and Tower Shield Experiments

Reactor experiments were used from the earliest days to support shielding studies and development of the necessary shielding properties data base. Prominent in this group were the Tank Shield Experiment with water shield base, and the Tower Shield Experiment, with air as the radiation transmission medium.

A good part of the data base for the design of radiation shielding, as well as the early methodology underlying the associated experimental work, was developed through the inspiration of the late Everitt P. Blizard. Working at Oak Ridge National Laboratory, he was instrumental in establishing there in 1948 the Lid Tank Shielding Facility (LTSF), to be followed one year later by the Bulk Shielding Reactor (BSR). A good part of the shielding information in use today as benchmarks for testing radiation shielding codes was derived from the experiments conducted at the BSR. This facility used a reactor immersed in a tank of water; hence its name. Various shield arrangements were placed against one face of this reactor, and the relevant shielding parameters were measured.

These facilities constituted the test bed for investigation of the principal interactions of importance to shield design. The experiments carried out there helped to establish the importance of secondary radiation sources, due to neutron capture in the shields. Shielding properties of various concretes were measured, and laminated water–metal–water shields were investigated.

To support a program of research and development of mobile reactors for aircraft and space propulsion, the Tower Shielding Facility (TSF) was constructed in 1954, and added to the experimental tools available to designers. TSF consists of a reactor and its associated shielding, supported in mid air by cables attached to 96 m high towers, located at the corners of a 30 m by 60 m rectangle. These dimensions were chosen to minimize the effects of back-scattering of radiation from nearby objects, and thus improve the quality of the shielding data obtained. TSF proved so convenient for experimental work in shielding that its operation was continued even after the ANP and the NERVA projects were terminated in 1961 and 1972, respectively.

Experiments simulating upper axial shielding for the Clinch River Breeder Reactor were done at TSF in the 1970s.[1]. Twelve decades of neutron flux attenuation were measured through shield mockups, consisting of various combinations of stainless steel, followed by sodium and carbon steel.

Since 1975, experiments in the TSF have supported development of the physics design of the Gas Cooled Fast Reactor (GCFR). A prototypic grid plate arrangement was tested with the aim of reducing radiation streaming through the gas coolant passages when aligned with spaces between fuel pins. In this series, the generic fuel pin streaming experiment was set up to measure neutron streaming in the axial direction of a triangular fuel pin lattice, using a collimated neutron beam originating in the reactor. To prevent the uncollided neutron beam from the driver reactor from becoming a dominant source of neutrons, a shadow shield arrangement was used between the reactor and the tested lattice. Experimental results were compared with Monte Carlo and discrete ordinates transport theory calculations set up both in discrete geometry and in the homogenized model. In the process, it was determined that the prediction of axial streaming in the fuel pins was easier than the accurate calculation of angular distribution of neutrons transmitted through the relatively large grid plate penetrations.[2]

Experiments for the determination of neutron transmission and gamma ray heating in the UO_2 and ThO_2 blanket and the inner and outer shields were also performed in the facility, with emphasis on different arrangements of stainless steel and borated graphite components[3].

REFERENCES

1. R. E. Maerker and F. J. Muckenthaler, "Measurements and Calculations of Neutron Fluxes Through a Simulation of the CRBR Upper Axial Shielding," *Proc.* Int. Conf. on Reactor Shielding, Knoxville, TN, CONF-770401-8, (18 Apr. 1977).
2. D. E. Bartine, D. T. Ingersoll, and C. O. Slater, "Recent Shielding Experiments at the ORNL Tower Shielding Facility," *Trans. ANS*, 33, 664–665 (Nov. 1979).
3. F. J. Muckenthaler, J. L. Hull, and J. J. Manning, "GCFR Radial Blanket and Shield Experiment," ORNL-TN-7237, Oak Ridge National Lab., Oak Ridge, TN (1980).

1.2 SOURCES OF FUEL THERMO-HYDRAULIC DATA

A considerable store of knowledge has accumulated from experimental work done in the study of piping and pressure vessel construction specific to the nuclear field, drawing on the large data base already existing in the chemical industry and steam power plant engineering. Flow loops were constructed by all principal reactor manufacturers and national laboratories, to study the various flow and heat transfer phenomena, the effect of surface corrosion and crud deposits, and the consequences of full flow blockage.

In the flow tests carried out with light water as heat transfer medium, three categories of tests stand out. They are the full length pressurized water tests, full length boiling water tests and scale tests. In the first two categories, simulated fuel rod arrangements in size up to a reactor fuel assembly are used with proper grid supports and assembly can. These assemblies are, as a rule, heavily instrumented for the measurement of temperature, pressure, flow, mechanical deflection and vibration effects.

One of the problems facing the designer of the thermo-hydraulic part of a reactor system is that of verifying experimentally the flow patterns and heat transfer uniformity of cores with a very large volume and flow passages numbering in the tens of thousands. One way to tackle the problem is to use scale tests. As a rule the scale tests use a smaller than full size fuel rod and assembly model, but attempt to simulate a part or a whole core loading pattern, including control rod thimbles or slots. Early in the development of nuclear reactors numerous thermo-hydraulic models of reactor systems were constructed. At first 1/8, 1/7, or 1/4 models were in use, and nuclear heating of the elements was simulated with electrical heating. Various flow regimes were studied in the scaled down reactor vessel, main coolant piping and steam generators. In some cases entire scaled flow systems of water reactors were constructed, including pumps, pressurizer, heat sink, etc., such as the Westinghouse PWR scale model built at the Forest Hills, PA facility.

Other facilities of this kind were erected with the specific objective of testing the scaling laws for the flow regime in a small test reactor, to give confidence in the validity of scaling up results from the small reactor to an industrial size power producer. A scale facility constructed for the support of the LOFT experiment is described in a later section.

Portable flow loops are also used occasionally to test the hydraulic resistance of full-size fuel elements in the field. Such test loops can be mounted on a trailer-truck and transported to the installation where the tests are to take place.

PWR FLECHT and BWR Refill-Reflood Test Loops

An important series of experiments in the area of PWR flow safety is the Full Length Emergency Cooling Heat Transfer test, the so called FLECHT series. Actively sponsored by the Electric Power Research Institute, it has produced a large body of experimental results and computer code verifications, and has led to the general firming up of an area considered very important to reactor performance and safety. These tests were carried out at the Westinghouse test facility in Forest Hills, Pittsburgh, PA.

In the FLECHT series 21 and 161 fuel rod bundles were placed in specially constructed flow loops and subjected to flow tests at variable heating rates. The coolant flow rate could be closely regulated, and variations in back pressure imposed to simulate normal and accidental core conditions. The nuclear heating of the fuel rods was simulated electrically. Free and blocked flow paths were tested, and overpower conditions could be overridden by forced, or gravity, reflooding of the bundle.[(1)–(3)] The range of test parameters used in FLECHT included inlet velocities from 1 to 45 cm/sec, pressure range of 0.1 to 0.6 MPa, and peak rod power from 1.7 to 4.6 kW/m with corresponding initial peak cladding temperatures of 140 to 1100^{0}C and inlet water subcooling of 70 to 90^{0}C. To reproduce the axial power distribution realistically, the electric heating of the rods was arranged to produce either a cosine or a skewed power distribution.[(4)–(6)]

A series of tests similar to FLECHT, but with thermo-hydraulic conditions suitably adjusted to boiling water reactor conditions, was sponsored by EPRI in support of the development of BWR safety technology. These tests are being carried out at the General Electric Co. Steam Sector Test Facility[(7)] (SSTF) at San Jose, CA. SSTF is a large scale facility designed to provide the experimental capability for the measurement of the various two-phase flow effects in mockups of single BWR fuel assemblies,[(8)] and for obtaining the system response data. Model assessment for BWR computer codes is also one of the aims. The experimental program[(9)] established for the facility is aimed at developing detailed experimental information and a realistic modeling capability for loss-of-coolant accidents in BWR. Of particular interest is the investigation of flow and heat transfer phenomena involving the refill-reflood process present in such hypothetical accidents. The experimental program of SSTF is well under way,[(10)] and the experimental results are being used in code assessment.[(11)]

The two series of tests described above provide a counterpart basis for reactor safety development, which also includes the LOFT experimental program. By concentrating attention on the thermo-hydraulic phenomena occurring internally to fuel bundles, they supply an independent check for the many thermo-hydraulic computer codes used in reactor design.

REFERENCES

1. "PWR FLECHT Separate Effects and System Effects Test (SEASET) Program Plan," RP959-1, Westinghouse Electric Co., Pittsburgh, PA (1978).
2. "PWR FLECHT SEASET 21-Rod Bundle Flow Blockage Task—Data and Analysis Report," EPRI NP-2014, 2 vols., Westinghouse Electric Co., Pittsburgh, PA (1982).

3. "PWR FLECHT SEASET Unblocked Bundle, Forced and Gravity Reflood Task—Data Evaluation and Analysis Report," EPRI NP-2013, Westinghouse Electric Co., Pittsburgh, PA (1982).
4. E. R. Rosal et al., "FLECHT Low Flooding Rate Cosine Test Series Data Report," WCAP-8651, Westinghouse Electric Co. (Dec.1975).
5. E. R. Rosal et al., "FLECHT Low Flooding Rate Skewed Test Series Data Report," WCAP-9108, Westinghouse Electric Co. (May 1977).
6. G. P. Lilly et al., "PWR FLECHT Skewed Profile Low Flooding Rate Test Series Evaluation Report," WCAP-9183, Westinghouse Electric Co. (Nov. 1977).
7. J. E. Barton, D. G. Schumacher, J. A. Findlay, and S. C. Caruso, "BWR Refill-Reflood Program 30^0 SSTF Facility Description," EPRI NP-1484, Electric Power Research Institute (Sept. 1982).
8. "BWR Refill-Reflood Program—Single Bundle Experimental Task Plan," 2 vols., EPRI NP-1524, General Electric Co. (1981).
9. G. W. Burnette, "BWR Refill-Reflood Program—Program Plan," EPRI NP-1522, Electric Power Research Institute (Oct. 1981).
10. "BWR Refill-Reflood Program—Core Spray Distribution Final Report," EPRI NP-1580, General Electric Co. (1981).
11. J. G. M. Andersen and K. H. Chu, "BWR Refill-Reflood Program Constitutive Correlations for Shear and Heat Transfer for the BWR Version of TRAC," EPRI NP-1582, Electric Power Research Institute (Jan. 1983).

Semiscale Facility for the LOFT Program

The initiation of the LOFT program, with a 50 MW(t) reactor simulating the behavior of a 3000 MW(t) commercial PWR, made it imperative to verify the scaling laws to be used. To provide this verification, a semiscale facility was constructed at INEL[1] and operated to obtain the required data. These tests check the validity of scaling laws applied to LOFT results, by pinning them down at the lower end, through the use of electrically heated core mockups, with 1.68 m long simulated LOFT fuel rods, as well as with the 3.66 m commercial fuel.

In the selection of the scaling approach to be used, emphasis was placed on true reproduction of the processes of primary concern in PWR loss-of-coolant accidents. These were transient saturated blowdowns involving two-phase flow. Of the possible choices, linear scaling preserves length-to-diameter ratios, but distorts the energy distribution process and results in a large reduction of the time scale. The classical dimensionless number approach was not understood well enough to apply it to complex, two-phase transient flow situations. For this reason, volumetric scaling proportional to thermal power was selected. The volume and power of the semiscale test article were scaled with respect to LOFT at approximately the same magnitude as LOFT was scaled to a commercial PWR (1/30 and 1/60, respectively). Where practicable, the ratios of volumes of the total system, components and pipes were preserved. For LOFT itself, the scaling is as follows:

- The linear heat generation rate of the large reactor is used.
- LOFT powers are scaled to large size as component volumes.
- Flow areas are scaled to provide identical mass fluxes.
- Component resistances are set identical to large reactor values.
- Pipe break areas are set in the ratio of core volumes.
- Pressure, temperature and mass flux are identical to large reactor values.

The results of the semiscale tests[(2)–(4)] show generally good agreement between the quantities scaled up and the experiment, although there has been some indication that the scaling rationale used leads to some distortion of the physical length. This is considered important only to phenomena dealing with subcooled decompression loading. Some distortion is also present in the ratio of surface area to fluid volume ratio. This is considered unimportant except in the analysis of effects taking place in the downcomer area of the core.

REFERENCES

1. L. J. Ball et al., "Semiscale Program Description," TREE-NUREG-1208 (July 1978).
2. H. S. Crapo et al., "Experimental Data Report for Semiscale MOD-1 Tests S-06-5, S-03-6, S-03-7, and S-03-8," ANCR-NUREG-1308 (1976).
3. B. L. Collins et al., "Experimental Data Report for Semiscale MOD-1 Test S-06-3 (LOFT Counterpart Test)," NUREG/CR-0251 (July 1978).
4. M. A. Lanzerman and D. S. Chu, "Effect of Scaling Compromises Between Semiscale MOD-1 and LOFT During Large Break Loss-of-Coolant Experiments," SEMI-TR-011 (July 1979).

1.3 DEVELOPMENT OF WATER REACTORS

Reactor experiments constitute a continuing source of data for the field of nuclear engineering. In these experiments, measurements on fuel rods, fuel assembly and reactor core are performed in a reactor operating in steady state conditions, or power transient conditions created artificially. By means of appropriate instrumentation or detailed post experimental examination of discharged fuel, data of great interest to the design engineers are obtained.

In the following sections, an overview of the power demonstration program will be given, followed by more detailed description of principal ongoing water reactor experiments; and the main references to reports containing the initial evaluation of the results will be quoted. Of particular interest here is the great effort connected with reactor experiments for the study of loss-of-coolant accidents in the LOFT project, and the Light Water Breeder Reactor project. Besides obtaining voluminous data on the behavior of reactor cores and emergency core cooling systems under accident conditions, the LOFT experiment has provided a verification of principal computer codes for reactor safety analysis.

Fission Reactor Experiments

In the program for the development of nuclear industry, construction of the lattice experiments and critical zero power assemblies were followed by the construction of experimental power reactors in gradually increasing size. Reactors built in the United States as part of that program were equipped with power conversion machinery, that is, turbines and generators, and approximated in their control requirements those of commercial, power producing reactors. They were not, however, operated under the same rules as the commercial power stations and could devote a substantial part of their life to the execution of experimental programs. Although several reactor types were constructed in this phase, water reactors predominated.

Table 1 lists power-generating experimental reactors in three categories, differing in their electrical power rating by approximately a factor of three. Only reactor types with commercial potential are included in the table. Reactors for space or defense applications were excluded. Name, reactor type, period of electric power generation, net electrical power and the cost of construction, when available, are given. The reactors listed were built and operated for a fixed period, as a rule much shorter than the usual 30-year life span predicted for a commercial power station. Experimental results obtained from the operation of these pilot plants made it possible to select the most promising reactor concepts for the next phase of power plant construction. Of the reactors listed, only EBR-II, Big Rock Point and La Crosse are still operated today.

Construction of power-generating experimental reactors was followed by the power demonstration program proper. This program was carried out in the 1960s and early 1970s with the objective of testing principal reactor concepts of relatively modest rating, in the range of 175 to 350 MW(e), in a commercial power plant setting. On the basis of generated operating data, the most promising concepts were then selected for further central power station development. Dresden 1 of 200 MW(e), Yankee-Rowe of 175 MW(e), and Fort St. Vrain of 330 MW(e) were built in that development phase. The last two plants are still operating. Clinch River Breeder Reactor Plant, at 350 MW(e), was also designed in the category of power demonstration reactors.

Numerous other reactor experiments were performed, particularly in connection with the space program. The results of these experiments

Table 1. Summary of U.S. Power Generating Reactor Experiments

Small 4–7 MW(e)		Medium 11–22 MW(e)		Large 40–75 MW(e)	
Saxton PW 1962–72	4 MW(e) M$ 6.5	Savannah PW 1962–71	17.6 MW(e)* (n.a.)	Shippingport PW 1956–82	50 MW(e) (n.a.)
Valecitos BW 1957–63	5 MW(e) M$ 3.1	BONUS BW 1964–68	16.5 MW(e) M$ 12.3	Humbolt Bay BW 1963–76	63 MW(e) M$ 24.3
Argonne BW 1956–67	5 MW(e) M$ 6.5	Elk River BW 1963–68	22 MW(e) M$ 13.5	La Crosse BW 1973–	50 MW(e) M$ 28.0
		Piqua OM 1963–66	11.4 MW(e) (n.a.)	Big Rock Pt. BW 1964–	63 MW(e) M$ 24.0
				Pathfinder BW 1966–67	59 MW(e) M$ 24.0
		SEFOR FR 1969–72	20 MW(e) (n.a.)	Enrico Fermi FR 1966–72	61 MW(e) M$ 26.0
St. Susana SG 1958–66	6 MW(e) M$ 8.8	EBR-II FB 1964–	16.5 MW(e) M$ 32.3	Hallam SG 1963–64	75 MW(e) M$ 55.4
		CVTR HW 1963–67	17 MW(e) (n.a.)	Peach Bottom GC 1967–74	40 MW(e) M$ 24.5

Reactor type: PW—pressurized water; BW—boiling water; FB—fast breeder; FR—fast reactor; SG—sodium graphite; GC—gas cooled; HW—heavy water; OM—organic moderated

* 22,000 shp

were documented in technical reports and are available for purchase to interested designers through the National Technical Information Service (NTIS) in Springfield, VA.

LOFT Experimental Program

High on the list of experiments dealing with reactor performance and safety is the Loss of Fluid Test (LOFT) series.[1] Conceived as a logical next step after a series of reactor transient tests, conducted at the Idaho Reactor Testing Grounds with the unpressurized core assembly (SPERT), it was intended to furnish more realistic data on the performance of a power reactor subjected to a series of tests with carefully graduated power levels. LOFT tests aim at reproduction of the various possible disturbances in the flow of coolant in a water reactor operating at full temperature and pressure. Simulation of pipe breaks, stuck valves and coolant leaks was desired.

The LOFT test article is a reactor of 50 MW(t) power rating, cooled and moderated with light water. The reactor core is about 1.7 m long, with an equivalent diameter of 0.6 m. It contains 1300 fuel rods with uranium enriched to 4.0% ^{235}U. The reactor is mounted on a platform on rails to make transfers possible from the containment building to the nearby "hot shop." Such transfer would be carried out in the event the reactor was damaged in the tests to the point of needing remote control devices for its postmortem examination. In actual practice, in all the tests conducted up to the time of this writing the damage to the reactor was minimal, and removal to the hot shop unnecessary.

The type of testing for which the LOFT facility is designed involves simulation of minor and major pipe breaks with or without loss of power to the pumps.[(2),(3)] Pipe breaks can be simulated through the opening of specially powered, quick opening, large blowdown valves, which permit completion of their action within 0.02 second. The system was shown capable of core reflooding within one minute after the given simulated break was observed. During experimental runs fuel cladding temperatures are monitored, pressures are measured, and the response of the Emergency Core Cooling System (ECCS) to the break is recorded in detail. The measurements made in the 50 MW(t) LOFT facility need scaling up by a factor of 60 before the results can be applied to full size commercial power plants of 3000 MW(t) rating. The Semiscale Facility at INEL provides verification of the scaling laws.

Construction of the LOFT facility was completed in late 1975. It started operating in 1976 with cold (nonnuclear) tests using a dummy core. Mini-blowdown experiments were started in January 1976 and nonnuclear large break experiments in March 1976. These tests were followed by hot (nuclear) tests of increasing severity. In 1979 the severity of the tests was increased, and major coolant pipe breaks were simulated at graduated power levels. The test program was continued into the 1980s with loss-of-coolant experiments at varied break size and power level and with operational transients, where specific components of support systems were operated at abnormal conditions. Altogether, between 1979 and 1982, twenty more different tests were carried out to simulate various loss-of-coolant accidents and operational transients. The experimental series was carried out with considerable involvement of reactor manufacturers, national laboratories engaged in reactor safety work, and the Electric Power Research Institute (EPRI). The main outlines of the experimental plan are summarized in Table 2.

Table 2. LOFT Experimental Program up to 1982

Year	Series	Description of Test Series
1976	L0	Ten mini-blowdown nonnuclear experiments to qualify components and operational procedures for simultaneous opening of quick-opening blowdown valves.
1976–78	L1	Six nonnuclear loss-of-coolant experiments performed at isothermal primary coolant conditions, and various simulated pipe break combinations.
1979	L2	Five nuclear loss-of-coolant experiments during ascension to full power of 52.5 kW/m, including simulation of double-ended cold leg large pipe breaks.
1979–80	L3	Three nuclear loss-of-coolant experiments to study the effects of small (6.4 cm) cold leg primary coolant pipe ruptures, and scaled pipe cracks.
1981	L5	One nuclear loss-of-coolant experiment with single-ended cold leg intermediate size pipe rupture, simulating the break of an accumulator pipe.
1980–82	L6	Fourteen experiments on anticipated transients with or without scram: control rod withdrawals, loss of feed water to steam generators and a boron dilution accident.
1980–81	L8	Two severe core transient experiments simulating core heatup in highly voided conditions, core uncovery and ultimately fuel damage with undisturbed core geometry.
1981–82	L9	Four anticipated transients without a control scram, including loss of off-site power and multiple system failures more severe than design basis accidents.

Note: Tests from two series were combined in some cases, when they could thus be carried out more expeditiously without interference.

The large volume of experimental data obtained in tests[4]–[11] is still undergoing analysis. To make the interpretation of these experimental results unambiguous, a considerable amount of effort went into the formulation of test methodology. In particular, an attempt was made to eliminate internal inconsistencies, define the precise meaning of the quantities measured, and specify measurement uncertainty.[12]

In a typical anticipated transient without scram (ATWS)[10],[11] test, the loss of off-site power was simulated to cause the shutdown of primary coolant pumps. Replacement of off-site power with power input from the emergency standby diesels and batteries was also tested under realistic conditions. The effects on the core included a rise in coolant temperature and a reduction of power generation rate in the fuel through the automatic operation of negative coefficients of reactivity. The resultant rise in system pressure was relieved by the automatic opening of the safety relief valves. The temperature, pressure and coolant flow patterns were recorded in great detail throughout the transient for post-experiment analysis.

In execution of the above tests, the multichannel analog signals derived from the nuclear, flow, pressure and temperature instrumentation mounted on the test article were converted to digital form and recorded on magnetic tapes. After review for completeness and proper tagging, the experimental results were reformatted and entered into a data bank. They are available for interested analysts from the Administrator of the NRC/DAE Data Bank, EG&G Idaho, Inc., P. O. Box 1625, Idaho Falls, ID 83415.

An important application of LOFT test results is in benchmarking of computer codes used in reactor safety studies.[13] Computer simulations of various accident sequences reproduced in LOFT include the loss-of-reactor-coolant accident following a large pipe break. The difficulty of mocking up such an accident was the cause of much concern in the past, and the LOFT program successfully tackled this problem. Three of the tests in the LOFT series were designated International Standard Problems, and three were U.S. Standard Problems, as shown in Table 3.

A comprehensive summary report[14] describes the 1976–1982 experimental series and compares experimental and calculated results obtained using the RELAP code in versions 4 and 5. Figure 1 shows the arrangement of LOFT flow loop components relative to the reactor vessel, as described in the summary report.

Although the major series of experiments planned for the LOFT facility

Table 3. LOFT Experiments Designated for Use in Standard Problems

Intnl.	U.S.	LOFT Test	Description
5	7	L1-4	Cold leg break with a scaled area equal to a commercial PWR full pipe area.
	10	L2-3	Cold leg break with a scaled area equal to a commercial PWR full pipe area.
9	*	L3-1	Cold leg single-ended break with a scaled area equal to the flow area in a 4-in. Schedule 160 pipe connected in the cold leg of a commercial PWR.
11		L3-6 L8-1	Cold leg single-ended break with a scaled area equal to the flow area in a 4-in. Schedule 160 pipe connected in the cold leg of a commercial PWR with primary coolant pumps left on.

* Also designated as an NRC Licensing Requirements Standard Problem.

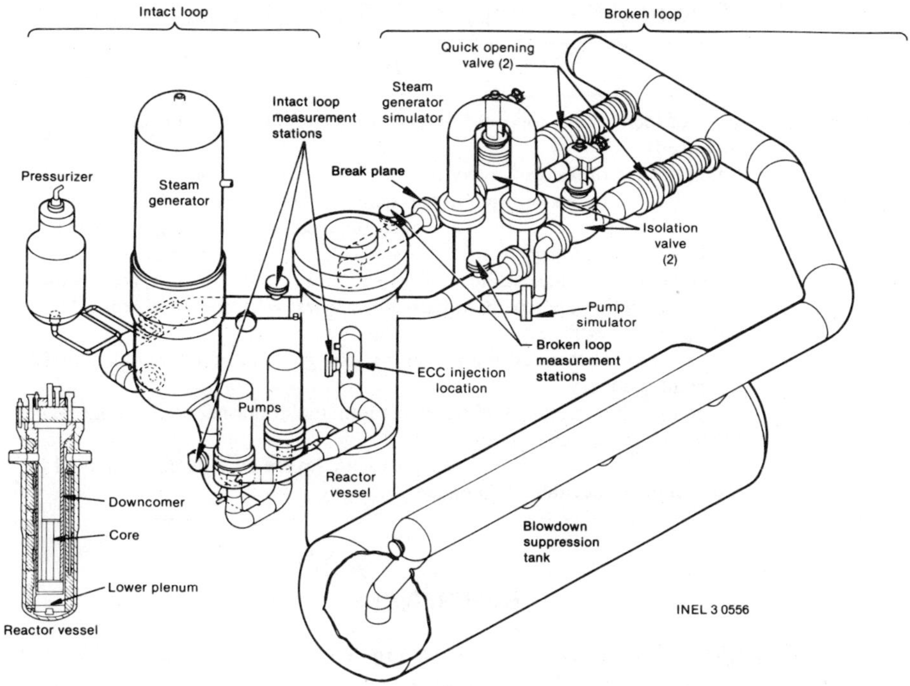

Figure 1. LOFT reactor and its flow loop components. (Reproduced courtesy EG&G Idaho, Inc.)

under NRC sponsorship came to an end in 1982, it was felt that much valuable additional information could be obtained by continuing the experiments with this well-established facility. An offer of continued funding was received from the Organization for Economic Cooperation and Development (OECD), and an extension of the experimental series was initiated in 1983. The funding for those tests was provided by Austria, Finland, Italy, Japan, Sweden, Switzerland, the United Kingdom and West Germany, with participation of NRC and EPRI. The experimental program for this phase of LOFT activity is outlined in Table 4.

The tests carried out in June and July of 1983, consist of simulation of loss-of-coolant accidents resulting from a break in the 7.6 cm diameter reactor outlet pipe with an active emergency core cooling system. The results of the first three of the above tests indicate that adequate reflooding and cooling of the core were maintained at all times by the emergency core cooling system.

Table 4. Joint U.S.– European LOFT Experimental Program

Test	Test Description
LP-FW-1	Loss of feed water, followed by primary system feed and bleed (pressurized fuel used).
LP-SB-1	Small hot leg break with early pump trip (pressurized fuel).
LP-SB-2	Small hot leg break with pump trip delayed 1 hour, (pressurized fuel used).
LP-02-6	Large cold leg break with pressurized fuel rods, loss of off-site power and minimum core cooling.
LP-LB-1	Large cold leg break with loss of off-site power and with emergency core cooling injection approximating assumed worst-case conditions for a United Kingdom power reactor.
LP-SB-3	Small cold leg break, core uncovery; steam generator feed and bleed used for plant recovery.
LP-FP-1	Fission product release and transport initiated at normal commercial power reactor operating conditions.
LP-FP-2	Fission product release and transport with severe core damage initiated by a risk-dominant accident sequence.

REFERENCES

1. D. L. Reeder, "LOFT System and Test Description," NUREG/CR-0247, TREE-1208, U.S. Nuclear Regulatory Commission (July 1978).
2. P. A. Harris, T. K. Samuels, and H. J. Welland "LOFT Experiment Operating Specification, Volume 2, Power Ascension Series L2," NE L2 Series, Rev. 2, EG&G Idaho, Inc. (July 1978).
3. J. D. Burtt, "Overview of the LOFT Experimental Program," Int. Colloq. on Irradiat. for Reactor Safety Programmes, Petten, Netherlands (June 1979).
4. M. McCormick-Barger, "Experiment Data Report for LOFT Power Ascension Test L2-2," NUREG/CR-0492, TREE-1322 (1979).
5. L. R. Feinauer, "Comparison of Experimental Results with Analytical Predictions for LOFT Test L2-2," EPRI-NP-1204, Electric Power Research Institute (Nov. 1979).
6. P. G. Prassinos, B. M Galusha, and D. B. Engelman, "Experiment Data Report for LOFT Power Ascension Experiment L2-3," NUREG/CR-0792, TREE-1326 (July 1979).
7. D. P. Bayless and M. J. Divine, "Experiment Data Report for LOFT Large Break Loss of Coolant Experiment L2-5," NUREG/CR-2826, EGG-2210, EG&G Idaho Inc. (Aug. 1982).
8. Donald B. Jarrell, Janice M. Divine, and Kathleen J. McKenna, "Experiment Data Report for LOFT Anticipated Transient Experiment Series L6-8," NUREG/CR-2930, EGG-2219, EG&G Idaho, Inc. (Nov.1982).
9. Paul D. Bayless and Janice M. Carpenter, "Experiment Data Report for LOFT Nuclear Small Break Experiment L3-6 and Severe Core Transient Experiment L8-1," NUREG/CR-1868, EGG-2075, EG&G Idaho, Inc. (Jan. 1981).
10. P. D. Bayless and J. M. Divine, "Experiment Data Report for LOFT Anticipated Transient Without Scram Experiment L9-3," NUREG/CR-2717, EGG-2195, EG&G Idaho, Inc. (May 1982).
11. Doyle L. Batt, Janice M. Divine, and Kathleen J. McKenna, "Experimental Data Re-

port for LOFT Anticipated Transient Without Scram Experiment L9-4," NUREG/CR-2978, EGG-2227, EG&G Idaho, Inc. (Nov. 1982).
12. G. D. Lassahan, "LOFT Experimental Measurements Uncertainty Analyses," Vol. 1, "Methodology and Summary," NUREG/CR-0169, EGG-2037, EG&G Idaho, Inc. (1983).
13. H. Ollikkala, "Post-Test Analysis of LOFT Anticipated Transient Experiments L6-1, L6-2, and L6-3 Using RELAP5," *Trans. ANS*, 41, 674–676 (1982).
14. Charles L. Nalezny, "Summary of Nuclear Regulatory Commission's LOFT Program Experiments," NUREG/CR-3214, EGG-2248, EG&G Idaho, Inc. (July 1983).

Shippingport Light Water Breeder Reactor

Also continuing, but in the terminal phase of operation, is the Light Water Breeder Experiment carried on at the Shippingport power reactor. Data on fissile fuel breeding in a water-moderated reactor are generated there in conditions resembling the conventional power station environment.

The Shippingport Light Water Breeder Project was advanced as a potential shortcut in the development of the breeder reactor. The objective of the project was to demonstrate the feasibility of fissile material breeding by replacing the regular low enriched uranium dioxide core in a pressurized water reactor with a core fueled with thorium. If successful, the concept would enable public utility companies to use lower-priced Th in place of enriched U, avoid generation of Pu in significant amounts, and produce discharged fuel of preferable composition from the point of view of non-proliferation of nuclear weapons. The conceptual designs of such power stations propose initial operation for some eight years in a pre-breeder phase, with the reactor fueled with U-235 and converting Th to U-233 in the blanket; followed by the asymptotic phase, during which the reactor operates on a three-year cycle on U-233 extracted from its blankets.

The Shippingport LWBR project aimed at a design breeding ratio of 1.01. Very tight lattices, combined with the seed-blanket modular core concept, and a very low volume fraction of structural material, Zircaloy-4, were used to increase the conversion of fertile to fissile material. The concept used Th-232 as fertile fuel, and generated U-233 through neutron capture in Th followed by the nuclear transmutation of Pa-233 into U-233.

To increase the fuel fraction in the core, it was composed of modules in the form of fissile seed islands placed in a sea of fertile material. Twelve such seed modules and twelve surrounding blanket sleeves constituted the center of the core. Its periphery consisted of 15 assemblies made up of large diameter ThO_2 rods clad in Zircaloy-4. The seed modules have a very low volume fraction of water, reducing parasitic neutron capture in hydrogen and sustaining a harder neutron spectrum, with corresponding enhancement of neutron production by the (n,2n) reaction and fast fission. The seed modules were suspended from control rod drives and were movable relative to the rest of the core. Each seed assembly had the fraction of U–Th variable along its length, allowing the insertion of additional fuel into the core.

The design of the LWBR core has been described in reports by Freeman, Connors et al. and Hecker,[(1)–(3)] whereas the results of physics tests and the associated experimental program were given by Sarber.[(4)–(5)] Hecker and Freeman have discussed the implications of the LWBR for fuel utilization in light water reactors.[(5)–(6)]

This project was in operation for five years, 1978–1982, using the plant originally designed for the Power Demonstration Project and started in 1957. The reactor was rated at 50 MW(e). Now that the five-year irradiation cycle has been completed, the fuel will undergo extensive isotopic analysis to ascertain the actual value of the breeding ratio achieved in the core.[(7),(8)]

Following reactor shutdown in October 1982, the core defueling schedule provided for shipment of 39 fuel modules to the Expended Core Facility at the Idaho National Engineering Laboratory, the shipments to be completed by September 1984.

At INEL, 500 fuel rods will be selected from 12 fuel modules and subjected to nondestructive fuel assay using irradiation of tested rods with a fixed neutron source, followed by scanning for the delayed neutrons emitted by the fission of U-233 contained in the rods. As an additional check, 17 fuel rods will be sent to Argonne National Laboratory for destructive assay and mass spectroscopic analyses. During 1985 through 1987 all fuel will be sent to INEL, and will be reprocessed and finally analyzed by mass spectroscopy for its isotopic composition.

REFERENCES

1. L. B. Freeman, ed., "The Calculational Model Used in the Analysis of Nuclear Performance of the Light Water Breeder Reactor (LWBR)," WAPD-TM-1314 (1978).
2. D. R. Connors, S. Milani, J. A. Fest, and R. Atherton, "Design of the Shippingport Light Water Breeder Reactor," WAPD-TM-1208 (1979).
3. H. C. Hecker, Jr., "Summary of the Nuclear Design and Performance of the Light Water Breeder Reactor (LWBR)," WAPD-TM-1326 (1979).
4. W. K. Sarber, ed., "Results of Initial Nuclear Tests in LWBR," WAPD-TM-1336 (1979).
5. W. K. Sarber, "Reactor Physics Experimental Program for the Light Water Breeder Reactor (LWBR) at Shippingport," WAPD-TM-1433 (1981).
6. H. C. Hecker and L. B. Freeman, "Design Features of the Light Water Breeder Reactor (LWBR) which Improve Fuel Utilization in Light Water Reactors," WAPD-TM-1409 (1981).
7. Louis B. Freeman and Herbert C. Hecker, "The Reactivity Worth of Protactinium-233 Inferred from Measurements, *Nucl. Sci. Eng.* 80, 338–341 (1982).

8. M. Natelson et al., "Nuclear Performance of the LWBR at Shippingport: 1977–1980," *Proc.* ANS Topical Meeting on Advances in Reactor Physics and Shielding, Sun Valley, ID (Sept. 14–19, 1980).

1.4 DEVELOPMENT OF FAST BREEDERS

TREAT Experiments

The Transient Reactor Test facility (TREAT) at Argonne National Laboratory has been engaged for more than a decade in an experimental program in support of the Liquid Metal Fast Breeder Reactor development program.[1] The focus of TREAT experiments has been the investigation of the behavior of fast reactor fuel subjected to transient overpower conditions. The power transients in TREAT are initiated by a rapid withdrawal of control rods, carried out at the rate of 1.8 m/sec. The movement of the rods controlling the transient can be stopped and rapidly reversed. In consequence, the reactor is capable of going in a "slow ramp" from 0 to 250 MW and again to zero power in less than 10 seconds. Energy releases in excess of 1 GJ are possible.[2]

The fuel testing volume of the TREAT reactor can accommodate entire fuel rods. In power pulsing tests of dry or gas cooled fuel, the test volume can be viewed through a slit in the moderator graphite. An arrangement of mirrors combined with a high speed camera permits filming of complete fuel response sequences, including fuel slumping and meltdown.

To accomplish the objectives of fuel testing in off-normal conditions, the TREAT reactor has been designed with a capability of subjecting the test sample to high neutron pulses of short duration. The graphite-moderated TREAT reactor has a large heat capacity in its moderator, giving it excellent power-quenching properties, which enable it to recover from power surges with relatively low heating of the core itself.

The reactor was recently upgraded[3] to extend its capability for testing larger fuel assemblies for the liquid metal fast breeder reactor under prototypical thermal-hydraulic conditions. The upgrade required provision of means for the reduction of damage to the reactor from "sodium hammer" effects, associated with burst-induced fuel vaporization events. Since then many fuel tests have been carried out with irradiated multipin bundles in which large spatial variations in coolant flow and power resulted in initial cladding failure in one of the pins located on the periphery of the fuel bundle.[4]

In recent tests[5] simulating the loss-of-flow accident sequence, three fuel elements containing $(Pu,U)O_2$ fuel were subjected to 20 to 30 times the normal power. The resulting fuel dispersal rates were measured and observed using a 1.2 m fast neutron hodoscope. The significance of fuel motion and its effect on reactivity could then be assessed by weighting the observed fuel density variations with distribution of reactivity worth typical for a liquid metal fast breeder reactor.

TREAT experimental results have been used to test the performance of thermo-hydraulic computer codes LEVITATE, SIMMER-II and COBRA-IV.

REFERENCES

1. C. E. Dickerman et al., "Status and Summary of TREAT In-Pile Experiments on LMFBR Response to Hypothetical Core Disruptive Accidents," *Symp.* Thermal and Hydraulic Aspects of Nuclear Reactor Safety, 2, 19 (Dec. 1977).
2. R. J. Page et al., "Results from the J1 Low Ramp Rate TREAT Experiment," *Trans. ANS*, 34, 545–546, Las Vegas, NV (June 9–12 1980).
3. S. K. Bhattacharyya et al., "Physics Design of the Upgraded TREAT Reactor," *Proc.* Conf. 1980 Advances in Reactor Physics and Shielding, pp. 144–155, Sun Valley, ID (Sept. 14–17, 1980).
4. A. B. Rothman et al., "TREAT Experiments With Irradiated Fuel Simulating Hypothetical Loss-of-Flow Accidents in Large LMFBR's," ANS/ENS International Meeting on Fast Reactor Safety Technology, Seattle, WA (Aug. 1979).
5. Richard Simms, E. Gerald, Alan B. Rothman, and George S. Stanford, "Loss-of-flow Transient Reactor Test Facility Tests L6 and L7 with Irradiated Liquid-Metal Fast Breeder Reactor Type Fuel," *Nucl. Technol.*, 52, No. 3, 331–341 (Mar. 1981).

Liquid Metal Flow Loops and Scale Tests

A separate category of experiments was created in the study of heat transfer to liquid metals due to the presence of problems specific to the interaction of liquid and solid metals present in the flow loops, particularly the problem of solubility of metals in each other, and the associated mass transfer. Of particular interest here was prevention of the conditions in which preferential leaching out of components of nickel-base alloys takes place from the high temperature regions in liquid-metal–cooled fast reactors and is subsequently followed by deposition in the lower temperature regions of the heat exchangers. This mechanism weakens the components in the high temperature region, and clogs up the narrow coolant passages in the colder regions.

For a good many years a substantial part of the development and testing of pumps and heat exchangers for liquid metals, as well as work on the interaction of liquid metals and reactor components, has been concentrated at Canoga Park, CA. Until 1978 Atomics International Co. operated the Liquid Metal Engineering Center there.[1] Properties of liquid metals relevant to cooling fast reactors were collected and disseminated by LMEC. Renamed the Energy Technology Engineering Center in 1979, and operated by Rockwell International, with a focus on fast reactor safety, the center is now equipped to test full-scale components of liquid metal reactor hardware. Components for FFTF and the Clinch River Breeder Reactor were tested there.[2] Piping design rules for high temperature liquid metals, based on Code Case 1592, were tested to confirm the validity of the underlying assumptions.

Of additional concern for the successful development of LMFBR technology is the ability to operate LMFBR fuel in the presence of a certain fraction of pins with cladding faults. The presence of these faults is acceptable only as long as fault propagation to adjacent pins can be prevented. To establish the conditions in which this requirement holds, an experimental program was instituted with the use of the Sodium Loop Safety Facility (SLSF), constructed and operated at the Idaho National Engineering Laboratory. As part of the LMFBR development program, SLSF supports a large sodium test loop. The test loop was inserted into the Engineering Test Reactor (ETR) and served as a test vehicle for seven tests of LMFBR fuel pins operating under simulated accident conditions. Double containment was used in the construction of the loop to minimize the possibility of sodium reaction with the ETR environment. In the last test of the series, molten fuel was injected into a 37-pin bundle of full-length fuel pins, with the objective of determining the severity of interaction between the molten fuel and the sodium coolant. The results of these tests indicate that failure propagation from pin to pin can be ruled out for oxide fuel, but flow blockage propagation cannot be ruled out and produces localized damage.[3],[4]

Scale models of the Clinch River Breeder Reactor were tested at the Stanford Research Institute,[5] Menlo Park, CA, to determine if reactor vessel integrity could be maintained under conditions of a hypothetical core disruptive accident (HCDA) involving the release of 661 MWs of energy. In these tests three 1/20-scale models of the reactor internals, vessel head and vessel itself, each mocked up with increasing degree of complexity, were instrumented to record pressures, strains and accelerations induced by the energy release. The upper internal structure, thermal liner and other details of construction were reproduced in the model.

REFERENCES

1. F. V. Naugle, "High-Temperature Piping-Design Technology, Jan.–March 1978," DOE/SF/76026, Atomics International Div., Canoga Park, CA, (1978).
2. M. J. Tessier, "Recent Developments in Liquid Metal Component Testing," *Proc.* 2nd Int. Conf. on Liq. Metal Technology, Richland, WA, pp. 1–9, (Apr. 20-24 ,1980).
3. R. W. Tilbrook, W. A. Ragland, D. H. Thompson, and S. S. Borys, "The Place of SLSF Experiment P-4 in Local Faults Evaluation," *Trans. ANS*, 44, 327 (June 1983).
4. J. W. Holland, "SLSF Experiment P-4 Post-Test Examination," *Trans. ANS*, 44, 333–334 (June 1983).

5. C. M. Romander and D. J. Cagliostro, "Structural Response of 1/20-Scale Models of the CRBR to a Simulated HCDA," *Nucl. Eng. Design*, 60, No. 2, 239–256 (Sept. 1980).

EBR-II, Fermi I and SEFOR Facilities

As part of the fast breeder reactor development plan an Experimental Breeder Reactor II was constructed in Idaho and went critical in 1963. The reactor is fueled with metallic uranium slugs, enriched to 67% in ^{235}U, clad in SS-304, and cooled with liquid sodium. It is rated at 16.5 MW(e), and is currently in operation, serving as a test bed for various experiments connected with development of fuels for fast breeders, as well as in reactor safety work.(1)

The Enrico Fermi I fast reactor was constructed in Michigan and went critical in 1963 as well as EBR-II, but had its operation terminated in 1972 due to coolant flow blockage, leading to a partial core meltdown. The reactor was rated at 61 MW(e), fueled by a uranium molybdenum (10%) alloy enriched to 25.6% in ^{235}U, and cooled with sodium. While it operated, it supplied valuable information on the performance of metal-cooled reactors.(1)

Both of the above-mentioned reactors operated with metallic fuels, but engineering studies indicated that ceramic fuels would give better performance, although several questions in the area of reactor physics remained to be clarified. In particular, early experimental resolution was required for the magnitude and sign of the Doppler reactivity coefficient of fast breeder reactors operating with oxide fuels, as well as the measurement of the effective delayed neutron fraction in cores with high plutonium content. This led to the decision to include in the fast breeder development plan the construction of a facility, designed specifically to test these effects.

The facility chosen for the task of measuring the fast breeder kinetic parameters was named the Southwest Experimental Fast Oxide Reactor (SEFOR).(2) Constructed at Fayetteville, AR, and started in 1969, the reactor had a power rating of 20 MW(t), and a core loaded with UO_2 and PuO_2 fuel, 18.7% enriched with fissile isotopes. It was equipped for testing fast reactivity excursion sequences and other reactivity effects.(3) After the experimental series lasting for over two years was completed, the reactor was shut down in 1972.

REFERENCES

1. M. Levenson, J. O. Arterburn, and W. H. Jens, "Major Contributions from EBR-II, SEFOR and Fermi," 4th Int. Conf. Peaceful Uses of Atomic Energy, Geneva, Switzerland (6 Sept. 1971).

2. "Southwest Experimental Fast Oxide Reactor (SEFOR), Facility Description and Safety Analysis Report," Amendment 30, Docket-50231-33, General Electric Co, Sunnyvale, CA (May 1969).
3. J. O. Arterburn, G. Billuris, and G. B. Kruger, "SEFOR Operating Experience," paper 71-NE-7, *Proc.* ASME Conf. Fast Reactors, Palo Alto, CA (7 Mar. 1971).

Fast Flux Test Facility

One of the major experimental programs currently in progress is concerned with the reliability of materials used in the construction of components for a fast breeder reactor. The facility established to provide an environment simulating that in a fast breeder reactor is the Fast Flux Test Facility (FFTF). It incorporates a liquid metal cooled fast reactor operating at high power density, but without breeding, and a large post-irradiation examination facility for the discharged fuel and components. The FFTF installation is located at the Hanford Engineering Development Laboratory.

The FFTF is an important second link in the development chain, which leads from the Experimental Breeder Reactor, through the FFTF and the Clinch River Breeder Reactor, to a full-scale commercial breeder reactor combining the function of fuel regeneration with a major power-producing capability.

The design of the reactor is based on the solid fuel concept and liquid sodium cooling. The fuel is in the form of mixed heavy metal oxide [PuO_2/$(Pu+U)O_2$], with fissile fraction varying from 22.4% to 27.4% by region, and with SS-316 fuel cladding.[(1),(2)] The thermal output of the reactor is 400 MW. There is no energy conversion plant attached to the reactor, and the heat generated is dissipated to the surroundings. The reactor went critical in February of 1980, and after a lengthy shakedown period, started its first major operating cycle in April 1982.

Operating cycles of 90 days duration are tailored to accommodate an extensive experimental program. Instrumentation placed in the core measures neutron flux levels, and core temperature and pressure, to ascertain the conditions to which samples of various alloys and materials are exposed. Performance of fuel assemblies of different design will be tested as well. Fuel burnups in excess of 100 GWD/MTHM are to be achieved in 1984.

In the design stage of the reactor, its mixed oxide fuel was tested through irradiation in the EBR-II reactor for fluences in excess of 9×10^{22} n/cm^2, E>0.1 MeV. During FFTF operation fluences an order of magnitude higher can be expected. Additional preoperational transient tests were performed on the FFTF fuel at Argonne and Hanford to assure its integrity.[(3),(4)]

Data on metal damage and failure rates due to interstitial gas formation and atom displacement from crystal lattices are among the results expected from this experimental series. Neutron damage to piping and core components is, of course, prominent among the safety and reliability concerns. In addition, the construction of the reactor gave focus to the development of components, and integration of component and system designs into an operating plant. Various licensing issues specific to reactors cooled with liquid metals were also clarified.

In addition to the data on performance of fuels and structural materials, the FFTF reactor supplies data on fast reactor physics. Of prime interest here are reactivity coefficients and feedback effects.[5],[6] These include the axial and radial fuel expansion, Doppler coefficient of reactivity, and sodium coolant density effects.

The Fuels and Materials Examination Facility (FMEF)[7] is located next to the FFTF reactor at Hanford, to provide the necessary laboratory space for detailed investigations of the irradiated materials and components. First examinations of spent fuel have already been initiated, and are expected to furnish much valuable materials data, of utility extending beyond fast breeder development into the area of interest to the fusion community.

REFERENCES

1. J. F. Williams and L. H. Rice, "Fuel Pin Fabrication for the FFTF," *Trans. ANS*, 34, 221–223 (1980).
2. L. V. Feigenbutz, "Instrumented Fuels Test for FFTF," *Trans. ANS*, 34, 247–248 (1980).
3. R. Simms, R. K. Lo, W. F. Murphy, G. S. Stanford, and A. B. Rothman, " Transient-Overpower Test E8 on FFTF-Type Low-Power Irradiated Fuel," ANL-77-93, Argonne National Lab. (1977).
4. F. E. Bard, B. C. Gneiting, W. P. Horn, T. R. Houser, E. J. Manthos, and E. H. Randklev, "Test Data and Post-Test Analyses of Three Prototypic-Length FFTF Fuel Pins Transient Tested in Flowing Sodium," HEDL-TME-79-9, Hanford Engineering Development Lab. (1979).
5. B. J. Knudson et al., "Reactivity Analysis of Core Distortion Effects in the FFTF," *Tans. ANS*, 36, 688 (1981).
6. R. A. Harris, B. J. Knudson, Yau-Ming Chien, and L. R. Campbell, "Reactivity Feedbacks in the FFTF: Prediction and Monitoring Techniques," *Trans. ANS*, 44, 518–519 (1983).
7. D. R. Doman, A. L. Trego, and C. A. Burgess, "Design Features and Objectives of the Fuels and Materials Examination Facility," *Trans. ANS*, 27,984–985 (1977).

Clinch River Breeder Reactor

Early in the development of the nuclear program, it became obvious that the enriched fuel technology selected for power reactors will utilize only a small fraction of the energy content of uranium. At the enrichment levels used currently, the mass of depleted uranium tailings from the gaseous diffusion separation plants exceeds by a factor of five the mass of uranium loaded as fuel into power reactors. Stored near the diffusion plants in a highly purified form as uranium hexafluoride (see Figure 2), this large mass of depleted uranium presently serves no useful purpose. Considering that environmental, mining, extraction and purification costs for this uranium already have been incurred, the question of its further use takes on added importance. The best use for this material would be as blanket material in breeder reactors. Considerations of this nature gave credence to early proposals that a breeder reactor development program be initiated. An orderly progression was visualized, involving design, experi-

Figure 2. Small part of depleted uranium tailings. (Reproduced courtesy Breeder Reactor Corporation)

mental and construction phases of increasing complexity and size, that would resolve the technical questions, provide operating experience, and protect the health and safety of the public.[(1)–(3)]

As the program evolved, the Experimental Breeder Reactor [20 MW(e)], Enrico Fermi I [61 MW(e)] and the Fast Flux Test Facility [400 MW(t)] were constructed, and extensive experimental programs connected with each of them were planned and carried out. Of these reactors only EBR-II and FFTF are currently in operation. The experimental programs connected with their operation were mentioned earlier in this guide.

The current step in this orderly developmental progression would be construction of the Clinch River Breeder Reactor, for which the design concept was selected in 1966 and the site picked two years later. By 1983 the design work was essentially complete, while ordering and construction of most components were also nearing completion. Of particular importance here was the design and testing of the J-shaped primary heat exchangers. Up until 1983 some $1.5 billion had been spent: $788 million for equipment completed and ordered, with $381 million worth of major components either in storage, or being tested.

The reactor was to be constructed at a total cost of $3.6 billion, with joint federal and private funding. The bulk of the development cost was to be borne by the federal government, and the bulk of the construction funds to be obtained from loans and bond issues. The construction of the reactor was halted in November of 1983 by congressional cancellation of counterpart funding.

From the engineering point of view, the reactor was to be a liquid-metal–cooled fast breeder rated at 380 MW(e), with a core region of 32.8% fissile enrichment and a breeding blanket loaded with depleted uranium. Once started, the reactor was to supply realistic operating data on its breeding capability, reliability and performance data on its components, and economic data on its operation. The last, and perhaps most important, aspect of its operation would have been be to provide a base for scaling up medium-size breeder reactors to the presently considered economic optimum of 1000 MW(e). Without the intermediate step provided by CRBR, scaling up to the commercial plant size will be difficult, unless the required data can be obtained from foreign reactor experience.

Early designs of the CRBR cores[(4)] predicted a breeding ratio of 1.20 at the beginning of core life and 1.23 at the end of the first cycle, dropping off to 1.12 in the equilibrium cycles. Later designs aimed at maintaining the value of 1.20 throughout the plant life. The design goal for fuel burnup was 150 GWD/MTHM. At peak linear fuel rating of 47.6 kW/m, a compound doubling time of 34 years was anticipated for the reactor.

REFERENCES

1. L. G. LeSage, "Fast Reactor Physics," *Trans. ANS*, 38, 7–11 (June 1981).
2. Donald R. Ferguson, "Fast Reactor Safety: Current Status and Future Challenges," *Trans. ANS*, 38, 24 (June 1981).
3. J. Costa, D. Grand, and R. Martin, "Fast Reactor Thermal Hydraulics," *Trans. ANS*, 38, 25–26 (June 1981).
4. J. A. Lake, R. A. Doncals, R. W. Rathburn, and H. C. Robinson, "Breeding Ratio and Doubling Time Characteristics of the Clinch River Breeder Reactor," *Proc.* Int. Conf. Advanced Reactors, Atlanta, GA, 665–576 (Sept. 1974).

1.5 DEVELOPMENT OF FUSION REACTORS

Magnetically Confined Fusion Experiments

The main thrust of experimental programs in fusion is not to prove that fusion is possible, but to show that the engineering problems that stand in the way of achieving plasma self-ignition, and subsequent energy extraction, can be solved.(1) High among them is the construction of a plasma-confining space, usually based on the magnetic bottle principle, and the design of energy-converting blankets.(2) The technology for magnetically confined fusion requires the use of high vacuum for plasma reaction chambers, superconducting magnets to obtain the most efficient use of magnetic fields, and advanced materials to provide components with better resistance to radiation damage. As these are developed, the effect of the fusion program extends far beyond its immediate area.

Several large experiments in controlled thermonuclear fusion are now under way, with several more in various stages of planning, design and construction. Their objective is to demonstrate that a self-sustaining fusion reaction can be achieved in conditions permitting the extraction of net output power. The key word here is "self-sustaining"; for although the energetics of fusion reactions are well known, and small neutron generators for producing 14 MeV neutrons are an off-the-shelf item in the laboratory supply trade, their mode of operation has a negative energy balance.

The physics parameters that must be met to achieve a self-igniting plasma are ion density, confinement time, plasma temperature and, in tokamak devices, the ratio of the strengths of poloidal to toroidal magnetic fields. For fusion reactions operating on the deuterium tritium fuel cycle, the bulk of the energy of the reaction is carried away from the site of the reaction by high energy neutrons. Extraction of this energy through interaction of neutrons produced in fusion with the materials of the fusion reactor blankets is associated with radiation damage.

Among the fusion experimental devices operating at the present time are the PLT, the Princeton Large Torus, and TFTR, the Tokamak Fusion Test Reactor at Princeton. Both radio frequency heating and neutral beam injection can be used in these devices.[(3)] TFTR generated its first light hydrogen plasma on December 24, 1982. The tests on the deuterium–tritium plasmas are not anticipated before 1986. The operation of the Poloidal Divertor Experiment (PDX) was initiated in 1979, also at Princeton, to permit studies of divertor configurations for the disposal of fusion debris. Next, four neutral beam injectors were installed to provide an additional 6 MW of auxiliary power. Plasma impurity control and plasma shaping are studied with this machine.

The Impurities Studies Experiment B (ISX-B) focuses on studies of reactor configurations with high poloidal to toroidal magnetic field strength ratio, plasma shaping experiments and study of the effect of plasma impurities on the operation of the device. In the early 1980s the installation of a bundle divertor gave additional capability to this device.

The ALCATOR reactors A and C, located at MIT in Cambridge, MA, are being used to study hydrogen and deuterium plasmas and the scaling laws governing plasma containment time, as well as the radio-frequency heating of plasma.

The DOUBLET-III machine of the General Atomics Laboratory at San Diego, CA, is operating to test doublet and D-shaped plasmas, as well as hydrogen injection into plasma using solid pellets. Since 1980, auxiliary heating of plasma has been provided with neutral beam injectors rated at 7.2 MW.[(4)]

EBT-S, the Elmo Bumpy Torus, is operating at ORNL with plasma heating with gyrotrons operating at 28 GH frequency. This is a scale experiment, used as a preliminary step to a larger machine. The tests carried out with it in late 1982 proved disappointing, with lower plasma temperature and shorter confinement time than anticipated, indicating that scale up to a larger machine may not be as successful as initially expected.

TMX is intended as a test of the plasma tandem mirror configuration rather than the toroidal configuration used in other machines. An electrostatic plug is being tested in a plasma configuration that does not turn back on itself. This experiment was intended to provide design data for the TMR device. TMR, the Tandem Mirror Reactor, is in the process of design and construction at the Lawrence Livermore National Laboratory in California.

The experiments mentioned above are preparing the groundwork for the design and development of fusion power reactors. In the process, extensive data are being obtained on materials performance in high radiation fields, cryogenic temperatures and very high magnetic fields. In this respect, fusion experiments share many of the concerns present in the development of fast breeder reactors, and in addressing them, contribute greatly to the development of materials science.

REFERENCES

1. "The Magnetic Fusion Energy Program—Its Objectives and Pace," Hearing Before the Subcommittee on Energy Research and Production, U.S. House of Representatives, 86th Congress, 1st Session, 11 Dec. 1979.
2. G. H. Miley, *Fusion Energy Conversion,* American Nuclear Society, LaGrange Park, Il (1976).
3. "TFTR Final Design Report," PPPL-1475, Princeton Plasma Physics Lab., Princeton, NJ (1978).
4. "DOUBLET III Results and Implications for Future Tokamaks," Doublet III and JAERI Physics Groups, General Atomics, *Trans. ANS*, 39, No. 2, 235–236, (1981).

Inertially Confined Fusion Experiments

The Lawrence Livermore National Laboratory is the site of the largest experiments in inertially confined fusion: the SHIVA and NOVA experiments.[1] The SHIVA experiment is based on multiple laser beam heating of small pellets of fusible mixtures of deuterium and tritium, performed at a rate high enough to produce inertial confinement of the plasma via the interaction of inward- and outward-directed shock waves.[2]

High power neodymium glass lasers were specially developed for use in inertially confined fusion.[3] An exceptionally high degree of synchronization and alignment of laser beams is required to focus the deposition of energy on a fuel pellet of almost microscopic dimensions. At high rates of energy deposition, an ablation of the outer layer of the pellet generates a shock wave. This, through an inertial effect, produces an implosion, heats and compresses the fuel in a range from 10 to 100 times the liquid DT density, and is capable of igniting fusion.[4]

The activation of a very powerful laser system is achieved by storing electrical energy in a high density capacitor bank capable of holding 25 MJ of energy. The laser beam optics are housed in a concrete canyon and supported by a rigid metal frame, required to achieve the necessary alignment and focusing of the laser beams. A 10 beam, 30 TW laser is available in the SHIVA configuration. A 400 kJ system of the 20 beam NOVA laser is also available. Both systems use automatic beam alignment,[5] a necessity in a region of frequent earthquakes.

REFERENCES

1. H. G. Ahlstrom, ''Laser Fusion Experiments, Facilities and Diagnostics at Lawrence Livermore Laboratory,'' UCRL-84002, Topical Meeting on Inertial Confinement Fusion, San Diego, CA (26 Feb. 1980).
2. D. R. Speck, E. S. Bliss, J. A. Glaze, B. C. Johnson, and K. R. Manes, ''Performance of SHIVA as a Laser Fusion Irradiation Facility,'' UCRL-82117, IEEE/OSA Conf. on Laser Engineering and Applications, Washington, DC (30 May 1979).
3. J. A. Glaze, ''SHIVA: A 30 Terawatt Glass Laser for Fusion Research,'' UCRL-80716, ANS Meeting, San Diego, CA (18 June 1978).
4. N. C. Holmes, R. J. Trainor, and R. A. Anderson, ''Laser-Generated Shock-Wave Experiments at Lawrence Livermore National Laboratory,'' UCRL-85625, Proc. AIR-APT Conf., Uppsala, Sweden (17 Aug. 1981).
5. R. D. Boyd, ''Evolution of SHIVA Laser Alignment Systems,'' UCRL-84008, Proc. Int. Symp. of the Soc. Photo-Optical Instr. Eng., San Diego CA (28 July 1980).

2. DATA ACCESS, STORAGE AND TRANSFER SYSTEMS

2.1 ORGANIZATIONS INVOLVED WITH OPERATING DATA

While in the past the industrial and national laboratories carried the bulk of design and development work on nuclear reactors, the task of operating and managing power generating nuclear installations fell, for the most part, to the private utility industry. The industry experienced the need for supporting organizations to orchestrate activities affecting all companies in that field, primarily in research, and in the collection and evaluation of operating data. In response to that need, three organizations were created by the users, and financially supported by their contributions.

The principal organizations created by the utility and reactor vendor industry are: the Atomic Industrial Forum, the Electric Power Research Institute and the Institute for Nuclear Power Operations. The activities and the type of information collected by these institutions in response to the industry's needs will be described below. Each contributes to the growth and digestion of the totality of operating, economic and technical information that constitutes the data base of nuclear engineering.

Each of these three organizations focuses its activities in a slightly different area, as indicated by their titles. Their activities, however, are broader than their names would indicate, a situation forced on them by recognition of the social concerns of the society, primarily through extension of their interests into the environmental protection area. The advocacy of nuclear energy in the political arena has been left to other organizations, created specifically for that purpose and supported by individual voluntary contributions.

Institute of Nuclear Power Operations

The accident at the Three Mile Island nuclear power station in the early hours of March 28, 1979, and the subsequent investigation of its circumstances by a presidential commission (the Kemeny Commission)[(1)] revealed the existence of weaknesses in the area of operator training and reactor operating procedures, and disclosed a lack of adequate resources within the utility company to respond to emergencies. The nuclear power industry responded to this challenging situation with the creation of the Institute of Nuclear Power Operations (INPO), charged with promoting safety in operations in nuclear power plants. In business since 1979, the Institute is a voluntary organization supported by all public utility companies in the United States that hold either a limited work authorization, a nuclear plant construction permit or a reactor operating license. The list of participants also includes those utilities that hold an equity position in a nuclear power plant, the nuclear supplier firms and some foreign nuclear organizations. The Institute is located at 1820 Water Place, Atlanta, GA 30339, tel.(404) 953-3600.

For operational purposes INPO is organized into the following divisions: administrative, analysis and engineering, communications, evaluation and assistance, radiological protection and emergency preparedness, training and education, international program. Its operating scope extends beyond the activities carried out at its Atlanta headquarters, and includes the support of various off-site training programs, nuclear plant inspections by visiting teams and sponsorship of seminars and symposia on topics of interest to reactor operating firms.

The resources of INPO are brought to bear on the operational side of the nuclear industry in two major ways: (a) through the direct assistance of visiting INPO personnel and by its (b) publications. The direct assistance consists of periodic evaluations of plant operating procedures by visiting teams, review and assistance visits in specific situations requiring immediate action, conferences to exchange information and, finally, observations of operating exercises performed to improve the emergency response of the operating personnel.

The publications of INPO provide the second way in which a distillation of operating experience, gained at many plants, is transmitted to all subscribers of INPO's support. Included here are descriptions of good practices, publication of the Emergency Resources Manual, preparation of materials for utility emergency planning and the operation of the NUCLEAR NETWORK communications system. Each of these activities is described in more detail below. In cooperation with the Electric Power Research Institute, INPO has developed and has managed since 1981 the Significant Event Evaluation and Information Network (SEE-IN), and publishes a quarterly *Review*, now in its third year.

Among the major efforts undertaken by INPO is an in-depth review of the Licensee Event Reports (LER). These reports are required by the Nuclear Regulatory Commission in each case of equipment malfunction, operator error, suspected tampering with the proper operation of equipment, or any other occurrence warranting further attention. The current level of submission of LERs runs at some 4500 per year.

Although the requirement to file the LERs with the NRC was intended to help the NRC in the execution of its statutory duties, analysis of the reports by INPO is thought to be an indispensable step in assuring that the industry will be able to police itself. Furthermore, the prompt and thorough analysis of events and sharing of the results of that analysis with all interested parties gives an additional impetus to efforts toward improving operator training courses, establishing safe and efficient operating procedures and preventing more serious equipment failures, where minor failures can be identified as warnings of more significant problems to come.

INPO experience shows that after analysis by reviewer teams composed of professionals with a strong commercial nuclear plant operating background, about 98% of LERs are found to be insignificant from the point of view of reactor safety. It is the remaining 2% that require full attention. Once an event report is judged significant, a Significant Event Report (SER) is transmitted through the NUCLEAR NETWORK, a communication system linking member participants of INPO. If further analysis and action are needed in the evaluating team's judgment, a Significant Operating Experience Report (SOER) is prepared and sent out to the participating members of INPO. Thus, one can think of the INPO Events Analysis Department as the extension of the operating groups at the plants. By bringing the experiences of all plants to one place, and sharing the results of their analyses, they greatly enhance the safety and reliability of nuclear plants throughout the nation.

To facilitate the speedy sharing of information between the participating utilities, the reactor manufacturers and the NRC, the Institute has sponsored the establishment of the NUCLEAR NETWORK communication system. This network uses a public telephone modulator-demodulator attachment and standard computer terminals to send and receive reactor operating information of interest to the participants. The items for transmission through the NUCLEAR NETWORK are selected by INPO's evaluation teams for their relevance to utility operations. About half of the members of the evaluation teams hold Senior Reactor Operator (SRO) licenses.

The NUCLEAR NETWORK works in conjunction with the Nuclear Plant Reliability Data System (NPRDS), established by the American National Standards Institute. NPRDS is a computerized data base containing information on the failure rate of systems and components in nuclear power plants. It has been in operation since the early 1970s, and also contains design and normal operating data. The system has a large memory storage capacity that, when combined with the NUCLEAR NETWORK, in effect creates a memory bank for the industry.(2) Thus INPO became an important contributor to the growth of the data base of the nuclear industry.

Another area of heavy INPO involvement has been improvement of the standards of reactor operator training, with particular stress on emergency preparedness. INPO review of existing operator training courses disclosed an emphasis on the normal operating mode of the reactor to the almost complete exclusion of training for emergencies. To remedy this shortcoming, which became painfully obvious after the TMI accident, training courses with the use of reactor simulators were revised, and a variety of accident situations set up to train the operators in dealing with them.[(3)] INPO also takes part in the specification and establishment of emergency response facilities,[(4)] with display and communication control centers set up at plant sites, reactor manufacturers', the NRC and the INPO headquarters in Atlanta, GA.

In the area of radiological protection, INPO compiled and distributed a Radiological Experience Notebook (REN) in 1981. To develop the radiological protection section for the General Employee Training (GET) guides, INPO held a workshop in the same year. The GET manual was issued for the first time in 1982. By participating in conferences, workshops and communication through the NUCLEAR NETWORK, INPO identifies material for inclusion in REN. Some input to this information medium comes from pertinent developments in the area of nuclear standards, and the NRC's regulatory guides. As new information is identified, loose-leaf supplements to REN are issued to enhance the training of reactor operating personnel in radiological protection, and promote good practice.

Although INPO is the youngest of the three organizations supporting the nuclear industry, it has already established itself as an important contributor to quality improvement in nuclear plant operation. Since 1982, INPO operating evaluation teams have been making inspection tours of nuclear power stations. The purpose of these inspections is to review the operating procedures of each plant in turn, and to suggest improvements based on the operating experience shared within the industry as a whole.[(5)] Taking into consideration that the membership of INPO now includes 105 utility companies and organizations involved in operating nuclear reactors, the sharing of their total experience can only be beneficial.

Electric Power Research Institute

To encourage cooperative support of research in areas of particular interest to the electric power industry, the Electric Research Council was organized in 1965. The Council consisted of representatives of the American Public Power Association, the National Rural Electric Cooperative

Association, the Tennessee Valley Authority and the Edison Electric Institute. One of its tasks was to assess the research effort required to match the projected expansion of the power industry in the forthcoming decades. The actual details of that task were entrusted in 1969 to the Research and Development Goals Task Force. Its report, issued in 1971, formulated utility research and development plans for the period up to the year 2000. During the period in question it is expected that the United States will require the replacement of worn-out generating equipment rated at some 200 GW(e), and an addition of some 300 GW(e) of new generating capacity. The size of this construction task is to be compared to the presently installed capacity of approximately 600 GW(e).

It was estimated that annual expenditures on the order of one billion dollars will be required from the utility industry, electric equipment manufacturers and the federal government for the supporting research. It was also determined that a new organization would be needed to handle this task. Such an organization, named the Electric Power Research Institute (EPRI), was established and started operations in 1973. It assumed the responsibility for management of R&D programs of the Edison Electric Institute, and succeeded the Electric Research Council. The headquarters for EPRI was established at 3412 Hillview Avenue, Palo Alto, CA 94303, tel. (415) 855-2411. Branch offices were established in Washington, DC, and in Santa Clara, CA, where a Software Center was organized. This last center moved in 1983 to Dallas, TX. It is operated as Electric Power Software Center by the University Computing Co., 1930 Hi Line Drive, Dallas, TX 75207, tel. (214) 655-8883. EPRI also owns and operates specialized research establishments at locations other than Palo Alto.

In its present form EPRI operates through the following divisions: Advanced Power Systems, Coal Combustion Systems, Electrical Systems, Energy Analysis and Environment, Energy Management and Utilization, Nuclear Power, Nuclear Safety Analysis Center (NSAC), Technical Information. The activities of the first of these divisions are heavily involved in research supporting the development of reactor safety, and thus contribute substantially to the formation of the data base of nuclear engineering.

In addition to sponsoring research, EPRI identifies areas requiring further study, channels industry research support to experimental and theoretical investigators, and, finally, prepares and disseminates information and experimental data from R&D activities in the field of electric power. An essential part of this activity is EPRI's large publication program of technical reports describing research carried out with EPRI's financial support. So far over 2500 technical reports have been issued by EPRI.[6] Of these, nearly half (1148) were sponsored by the Nuclear Power Divi-

Table 1. Output of EPRI Reports on Nuclear Reactor Technology

Number	Sponsoring Department	Number	Sponsoring Department
171	Engineering and Operations	103	Nuclear Power
481	Safety and Analysis	238	Systems and Materials
22	Miscellaneous Reports	30	Special Reports
46	Nuclear Safety Analysis Center		

sion and the Nuclear Safety Analysis Center. Table 1 summarizes the publication activity in various sections of these divisions, and highlights more important publications. In addition, brochures, pamphlets and books are published by EPRI, as a part of the public awareness program, and a monthly *EPRI Journal* is distributed to member organizations and interested parties. Its April issue contains the annual report.

To centralize information on research projects carried out by EPRI, the electric utilities and other research organizations dealing with the electric power industry, and to avoid duplication of effort, the Electric Power Data Base (EPD/RDIS) was established by EPRI. Currently it contains about 9000 records, of which 4000 originated with projects sponsored by EPRI, both completed and ongoing. A good proportion of them deal with matters related to nuclear applications. The data base content represents research activity supported at an annual industrial funding level of $500 million.

The records contained in EPD/RDIS consist of the project title, contractor, funding levels for past, current and future years, project description, identification of area of research through subject headings and classifications, and the list of resulting publications. On-line access to this data base is available to EPRI members and DOE contractors equipped with interactive terminals through the Dialog Information Services and the DOE/RECON network. An extension of access to this data base through other commercial retrieval services is contemplated for the near future. Access to EPD/RDIS is of particular value to electric utilities, architect-engineers, consulting firms and manufacturers, or any other organization concerned with the use, production or distribution of electric energy. The photocomposed hard copy of the current content of this data base is issued annually as "Digest of Research."

Three other data bases supported by EPRI are of interest to nuclear engineering. They are the Equipment Qualification Data Bank (EQDB) established with EPRI support by the NUS Corporation, the Fuel Performance Data Base (FPDB) and the Reactor Vessel Surveillance (RVS) data base. EQDB is accessible remotely, and is a computer-based interactive system for storage and retrieval of environmental qualification data in summary form for safety-related electrical equipment installed in nuclear plants, operating or under construction. The data on several thousand models of electrical equipment are organized in approximately 40 information fields, and originate from plant data submitted to the Nuclear Regulatory Commission by utilities and equipment manufacturers. Designers specifying qualified components, utilities carrying out qualification testing and plant operators searching for replacement equipment will find access to this data base invaluable.

FPDB, the Fuel Performance Data Base, contains information from ten power reactors and three test reactors concerning material performance of reactor fuel. Entered for 700 variables, it has currently over 300 megabytes of information. This data base consists of fuel performance data generated in EPRI funded projects, as well as data made available to EPRI on an exchange basis and restricted in disclosure to holders of an EPRI license. Even for data generated under EPRI funding, their release is under control of the responsible project manager. Application for access should be made through this channel. The mechanics of retrieval of information from this particular data base were made easier with the Relational Information Management System (RIMS), a relational interactive computer program permitting access to several data bases simultaneously during a single search.

RVS, the Reactor Vessel Surveillance data base, contains computerized data on 50 reactor units and 73 surveillance reports covering the majority of the PWR systems constructed in the United States. Information contained in this data base is of particular interest to personnel of safety and licensing organizations concerned with reactor vessel integrity.

The sponsorship by EPRI of the development of numerical methods in the nuclear field is adding significantly to the software inventory available. By selecting and developing computer codes, EPRI renders a substantial service to electric utilities handling their own fuel cycle and licensing activities, and helps them in the operation of their nuclear power plants. A list of codes presently available is given in Table 2.

Software packages for codes listed in Table 2 are available under EPRI license from the University Computing Co. The packages consist of the documentation, source code, input and output for test cases. The annual membership contributions to the EPRI organization cover the distribution fees for the software. The codes can also be used at the University Computing Co. as part of a service provided by EPRI to its members. In this case, input data can be transmitted from the utility to UCC either as cards

Table 2. Nuclear Computer Codes Available from EPRI

Code Name	Code Function
ABAQUS-ND	Finite element, nonlinear dynamic analysis
ARMP	Advanced fuel recycle methodology
BIGIF2	Fracture analysis in multidimensional stress fields
CASMO-1	Multigroup, 2-D transport and burnup calculation
CAT	Construction of fault trees from decision tables
COBRA	Analysis of reactor core thermal hydraulics
COMETHE	Prediction of fuel pin thermo-mechanical behavior
COPHIN	Transformation of CPM or CASMO output for input to PDQ
CPM	Multigroup, 2-D transport burnup in fuel rods
DATATRAN	Data base management and code execution control
DUVAL	Steam generator 2-D, two-phase, fluid flow
EPRI-CELL	Nuclide accretion and depletion in a fuel cell
FRAY	Steady state and transient fuel rod evaluation
GLASS-2	Global–local, finite element fracture mechanics
GO-2	Probabilistic system availability and reliability
HITCH	Composition and pH estimates of solutions, sludges
MANAGE	Multicycle fuel management
MEKIN	3-D reactor core kinetics
NODEP-2	Nodal 3-D steady state core power distribution
NORGE	Cross-section processor for SIMULATE code
PDQ-7/HARMONY	3-D neutron diffusion, with HARMONY burnup module
PHAMAS	Phased mission reliability analysis of systems
PSMS	BWR hybrid power shape monitoring system
RASP(BWR)	BWR neutronic and thermal-hydraulic simulation
RASP(PWR)	PWR neutronic and thermal-hydraulic simulation
RETRAN-02	Transient thermal-hydraulics of 1-D flow systems
SAMPAC-1	Steam bubble dynamics in subcooled pool
SIMULATE-E	LWR power distribution in 3-D by nodal solution
STEALTH	Structural, thermo-hydraulic and seismic analysis
TORMIS-2	Plant safety from tornado-generated missiles
VIPRE-01	Thermal-hydraulic analysis in 3-D for LWRs
WAMBAM	Probabilistic fault tree evaluation

by mail or as data files sent by wire, and the computer usage fees, incurred in program execution, are charged to EPRI members at a discounted rate.

Atomic Industrial Forum

The Atomic Industrial Forum is an organization established in 1953 in Washington, DC, to represent entities interested in the promotion and application of nuclear energy for peaceful purposes. AIF is incorporated in the State of New York as a nonprofit trade association. It provides a framework for exchange of thoughts and ideas among the executives of the utility industry and reactor suppliers on issues concerning management and technical problems. Some legal questions are also investigated.[7] At present AIF has 575 corporate and institutional members. Although the bulk of its membership comes from the United States (85%), 27 other countries are also represented. As a rule, the organization arranges to hold its meetings at the same time and place as the American Nuclear Society, to make it possible for the captains of the industry to participate in both functions.

The activities of AIF have dealt recently with topics concerned with the achievement of a more effective and speedier plant licensing process. As a part of this effort, AIF has drafted a proposed licensing reform bill to permit replacement of the present multistage licensing process with a one-stage procedure. This can be carried out within the existing legislative framework enacted by Congress for the establishment of the Nuclear Regulatory Commission.

In the legal area, AIF has filed briefs in support of a petition to the Supreme Court, requesting a ruling on the constitutionality of a nuclear plant certification law of the State of California. The law suspended the certification of new nuclear plants until the Energy Commission of the State of California is satisfied that a federally approved method of radioactive waste disposal has been developed.

The emergency planning activities stimulated by the accident at the Three Mile Island plant also resulted in AIF involvement. In digesting the lessons learned from that experience, AIF sponsored the Industry Degraded Core (IDCOR) rule-making program.[8] The aim of the program is to study costs and benefits of reactor design features added to deal with consequences of severe accidents. The program is funded by some 60 organizations, including utilities, reactor manufacturers and architect-engineer companies.

As part of post-TMI activity, AIF also sponsored a reexamination of the so-called Source Term, used to describe the quantity of an anticipated release of volatile radioactive material in reactor accidents. Experience gained from the TMI accident indicates that in the past the magnitude of the Source Term was greatly overestimated. The proposed distribution to the public of potassium iodide, to serve as a blocking agent against absorption by the thyroid gland of released iodine, may not be necessary, since the release of iodine at the TMI plant was minimal.

AIF has also reexamined, and presented to the NRC, industry views on the subject of qualification of personnel, and the transportation procedures to be used in transferring nuclear fuel, both new and discharged, from site to site. Security clearances, the frequency of en route position reports and supporting security arrangements were examined.

The viability of the uranium mining and processing industry has been studied by AIF with regard to the current state of the economy and the availability of enriching services, both at home and abroad. AIF participated, on industry's behalf, in discussions with the federal government concerning alternatives to modernization of uranium-enriching plants, and building additional enriching capacity using the centrifuge and laser-energized ionic methods. The assurance of fuel supply to foreign customers buying U.S. designed and fabricated reactors was also examined. Of particular concern here were the steps that could be taken to prevent the diversion of fissile materials from peaceful applications.

The participation of reactor and fuel manufacturing companies in the construction and operation of fuel reprocessing plants has been greatly slowed by delays in licensing of the Barnwell reprocessing installation. A review of the government's position is required to quell the industry's fears of making a bad investment. In the past, the federal government has been using AIF as a channel for its efforts to encourage increased financial participation of the nuclear industry in the completion of the Clinch River Breeder Reactor. The construction of a breeder plant would provide for the eventual use of the large stocks of depleted uranium, left as a byproduct of enrichment operations.

AIF has been following research developments in the field of controlled fusion and is interested in defining in advance the problems that will face the industry when the engineering of power producing fusion reactors commences. From time to time special conferences and workshops are held on subjects of topical interest to AIF membership, and proceedings are published.[9] The objective of these activities sponsored by AIF is to promote an exchange of information within the nuclear industry, universities and government. In addition to holding meetings and presentations on subjects dealing with the economics, planning and management of nuclear industrial facilities, AIF has since 1964 supported two monthly publications: a newsletter, ''AIF INFO,'' and a monthly journal, *Nuclear Industry*. An annual ''Nuclear Power Facts and Figures'' summarizes significant developments in the industry.

Since 1975 AIF has participated in the ''National Environmental Studies Project'' (NESP). As part of this project, outside contractors investigate and report on topics of current interest to the nuclear industry. The resulting series of reports[10],[11] is designated by AIF/NESP numbers. So far 26 reports have been published, varying in subject matter from record keeping on radiation exposure of transient workers in nuclear power plants, to studies of nuclear power plant siting and decommissioning procedures.

Another product of NESP is the creation by AIF of the EPIA data base. Formerly called INFORUM, this data base provides access to the relatively inaccessible literature on the environmental aspects of new power plants. Over 20,000 bibliographic citations cover among other things such information as power plant siting methodologies, licensing and permit data, safety and risk assessments, uranium mining and milling facilities, uranium enrichment and waste disposal facilities. All aspects of planning, construction and operation of electric power plants are covered through reports generated by the industry as well as the Nuclear Regulatory Commission, Department of Energy, Bureau of Land Management, Environmental Protection Agency and other federal and state agencies. Journal articles, conference proceedings and congressional hearing records are also included in this data base. The EPIA data base is available for searching on the ORBIT data base access service, as described later in this chapter.

AIF also publishes a series of short eight-page booklets informing the public on the application of nuclear energy, develops films and other audiovisual aids and in general conducts public information services. For additional information contact: Atomic Industrial Forum, Inc., 7101 Wisconsin Ave., Bethesda, MD 20814, tel.(301) 654-9260.

REFERENCES

1. "Report of the President's Commission on the Accident at Three Mile Island," John G. Kemeny, Chairman (Oct. 1979).
2. Bruce W. Logan, "SEE-IN, Notepad, and NPRDS: A Utility's Viewpoint," *Trans. ANS*, 44, 491–492 (June 1973).
3. "Accident Sequences for Design Validation and Training, Safety Parameter Display System," NSAC-40, Nuclear Safety Analysis Center (Apr. 1982).
4. "Functional Criteria for Emergency Response Facilities," NUREG-0696, U.S. Nuclear Regulatory Commission (1981).
5. "Performance Objectives and Criteria for Plant Evaluations," INPO-83-014, Institute of Nuclear Power Operations, Atlanta GA (1983).
6. "EPRI Guide, Winter 1982–1983," Electric Power Research Institute, Palo Alto, CA (1983).
7. Arthur W. Murphy and D. Bruce LaPierre, "Nuclear 'Moratorium' Legislation in the States and the Supremacy Clause: A Case of Express Preemption." Atomic Industrial Forum, Washington, DC (1975).
8. "Industry Degraded Core Rulemaking (IDCOR) Program," IDCOR Report 81-012-01, Technology for Energy Corporation (Oct. 1981).
9. "Electric Imperative," *Proc.* of an AIF Workshop, 14–17 June 1981, Monterey CA, Atomic Industrial Forum, Washington, DC (Mar. 1981).

10. Stone and Webster Engineering Corporation, "Analysis of Nuclear Power Reactor Decommissioning Costs," Atomic Industrial Forum Report AIF/NESP-021 (1981).
11. Dames & Moore, Inc., "Alternative Siting Requirements and Practices for Nuclear Power Plants," Atomic Industrial Forum Report AIF/NESP-018 (1980).

2.2 DATA BASES RELEVANT TO NUCLEAR INDUSTRY

To the extent that the information generated by the nuclear industry is computerized and made available, access to it is facilitated by the rise of organizations specializing in information transfer. Data base operating services and communication networks are the operating link between the computerized data base and the user. By introducing uniform search protocols, arranging for the use of dedicated communication lines connecting major cities, and, last but not least, handling the billing and bookkeeping of data base on-line searching, these organizations provide a valuable service.

The three data base services most important to nuclear engineering are: DIALOG, BRS and ORBIT. Each of them rents tapes containing data base materials from the organizations developing them, transcribes the tapes on high speed magnetic disks, and connects disk units to a central processor that handles access protocol programs and billing procedures. The same central processor handles interface operations with the TYMNET, TELENET and UNINET telecommunication networks that offer long distance, nationwide transmission lines operating at 300 and 1200 baud rate (30 and 120 characters per second).

Given access to a computer terminal, or a personal computer equipped with an asynchronous communications adapter plus a modem (modulator-demodulator) telephone coupler, the interested nuclear engineer is faced with choosing the data base access service. Selection depends, of course, on the choice of data bases offered, as well as on the service fees. Since some of the data bases of interest are offered by more than one access service, the final selection may depend on other factors, such as individual preference for a command-driven rather than menu-driven interrogation mode, or some other feature, such as the ability to order documents while on-line to the data base, or the availability of electronic mail service. A comparison of the access interrogation mode and main features of the three commercial information services is given in Table 3.

The DIALOG Data Base Service

One of the largest vendors of on-line services in the United States is the DIALOG service of California, in operation since 1972. Currently it offers access to more than 150 data bases, with a total of over 60 million refer-

Table 3. Features of Public Access Data Base Services

Feature	DIALOG	BRS	ORBIT
Interrogation by:	command	menu	command
Ordering on-line	yes	no	yes
Electronic mail	no	yes	no
Cost display	yes	no	yes

ences and abstracts in all fields of information. Among others, it carries the current updated catalog of all technical reports available for purchase from the National Technical Information Service. To access these data bases, the services of TELENET and TYMNET may be utilized at a user fee of $ 10 per hr, and 30 or 120 characters per second transmission rates can be used. Of the data bases available from DIALOG the following files are of particular interest to nuclear engineering:

NTIS — Catalog listing unclassified, technical reports available for sale to the public from NASA, DDC, HEW, DOE, DOT, HUD, Dept. of Commerce and other government agencies and their contractors. It contains 972,400 citations; covers 1964 to present. Connect charge is $45/hr.

COMPENDEX — A machine-readable version of the Engineering Index, covering approximately 3500 journals and monograph publications of engineering societies and organizations. It contains 1,174,000 records with monthly updates; Covers 1970 to present. Connect charge is $98/hr, plus 30c per full record typed or displayed on-line.

ISMEC — Index of significant articles in all aspects of mechanical, nuclear, electrical and electronic engineering, plus energy and power generation. It contains 145,900 citations; covers 1973 to present. Connect charge is $73/hr, plus 15c per full record printed or displayed on-line.

FRA — Federal Register Abstracts; gives a comprehensive coverage of federal regulatory agency actions, as published in the Federal Register. It covers 1977 to present; now contains over 184,000 citations. Connect charge is $75/hr.

ENERGY — The data base of DOE, containing references for all aspects of energy, including "Nuclear Science Abstracts" and "Atomindex." Available to U.S. users only, it now contains over 1 million records; covers 1974 to present. Connect charge is $40/hr.

Figure 3. Magnetic disk storage units of a large data base. (Reproduced courtesy DIALOG Information Services)

A subset of its data base access service is offered by DIALOG under the designation "Knowledge Index," for use at reduced rates in the evening and at night.

For additional information contact: DIALOG Information Services, Inc., 3460 Hillview Avenue, Palo Alto, CA 94304, tel. (800) 227-1927; accessed through TYMNET, TELENET, UNINET.

BRS, Bibliographic Retrieval Services

The name of this organization belies the broad range of its data base offerings, some of which include data bases of interest to the nuclear engineering community. Created in May of 1976, since 1980 BRS has been owned by Indian Head Inc. The distinguishing feature of this data base service is that it offers access to some full text data bases, such as those developed by the American Chemical Society.

Five subscription access plans for a committed number of search hours, and an open access plan are offered. Under the "open access plan" the data base access service is offered by BRS during business hours at a rate of $35/hour, plus data base royalties and communications network charges, which at the time of this writing were: UNINET $6/hr, TELENET $7/hr, TYMNET $11/hr. A sign-up fee of $50, applicable to data base searching done during the first three months of service, is also charged to cover the cost of setting up the customer account, and for providing the user code number. Of more than 70 data bases offered for access by BRS, those of particular interest to nuclear engineers and their royalty fees are listed below.

NTIS Prepared by National Technical Information Service, contains bibliographic information and abstracts of reports in all areas, including nuclear energy, mostly describing government sponsored work. Royalty: $8/hr.

COMP COMPENDEX, otherwise known as the Engineering Index; prepared by Engineering Information, Inc., contains bibliographic information on engineering reports of all types, including the nuclear area. Royalty: $38/hr.

INSP Produced by the Institute of Electrical Engineers, London, England, contains bibliographic information on engineering, physics and computer science. Royalty $35/hr.

SMIE Smithsonian Science Information Exchange; contains bibliographic information in the physical, social, engineering and life sciences. Royalty $30/hr.

"After Dark" service, for access to a subset of BRS data bases, is offered for on-line searching at reduced rates between 6 p.m. and midnight. A monthly user bulletin is published in two versions: the full service and the "BRS After Dark" service.[(1),(2)]

For additional information contact: Bibliographic Retrieval Services, Inc. (BRS), 1200 Route 7, Latham, NY 12110, tel. (800) 833-4707; accessed through TYMNET, TELENET and UNINET.

REFERENCES

1. *BRS Bulletin*, Vol. 7, No. 2, (Febr. 1983).
2. *After Dark Datalines*, Vol. 1, No. 1 (1983)

ORBIT Information Retrieval System

This data base access system was developed by a subsidiary of the Burroughs Corporation, the System Development Corporation. It offered in

1965 the first on-line bibliographic retrieval system. In existence since 1974 as a separate division, it now offers for on-line access over 70 data bases containing approximately 55 million abstracts and citations. Among the data bases offered for access, the following are of particular interest to nuclear engineers:

EDB	DOE Energy Data Base; a collection of bibliographic citations and abstracts of the world's scientific and technical literature on energy, covering, among others, nuclear fuels, nuclear power plants, nuclear reactor technology, fusion energy. Includes a full file of "Atomindex," and contains over 1 million records updated every two weeks. Prepared by the DOE Technical Information Center. Connect fee: $45/hr, plus 10c per citation or bibliographic record printed on-line.
COMPENDEX	The primary data base for engineers. Includes the Engineering Index, with many references to primary nuclear engineering sources. Connect fee: $98/hr, plus 25c per citation or bibliographic record printed on-line.
EPIA	Electric Power Industry Abstracts, formerly called INFORUM, produced by the Atomic Industrial Forum. Covers plant siting methodologies, licensing data, safety and risk assessment, uranium mining and milling, uranium enrichment, waste disposal facilities. Contains over 20,000 bibliographic citations from 1975 on. Connect fee $85/hr, plus 10c per citation or bibliographic record printed on-line.
POWER	Contains complete cataloging records of the Department of Energy Library devoted to energy, including former AEC, ERDA and the Federal Power Commission, as well as the Federal Energy Administration. Approximately 3000 titles of all nonserial publications of interest to DOE are added annually. Covers 28,000 citations from 1950 on; updated every two months. Connect fee: $40/hr plus 10c per citation printed.

There is no entrance fee required for becoming a user of the ORBIT system, but a "User Manual" and "Quick Reference Guide," offered for a total cost of $57, are required, together with a code access name and an account number. Description manuals of individual data bases are available for $7.50. Connect costs consist of data base connect fee, varying from $40 to $125 per hour depending on the data base, print fee between

10c and $3 per full format on-line citation, plus an $8/hr fee for TELENET or TYMNET line use. Offline printing and mailing of search results from some data bases is also offered. A bimonthly newsletter, "Searchlight," is published for subscribers.

For additional information contact: ORBIT Information Retrieval System, SDC Search Service, 2500 Colorado Avenue, Santa Monica, CA 90406, tel. (800) 421-7229; accessed through TYMNET and TELENET.

Oak Ridge Technical Information Center

The Technical Information Center (TIC) was created in Oak Ridge, TN, in 1946 by the Atomic Energy Commission to serve as a collection, processing and distribution point for nuclear information.(1) It operated in the same capacity under ERDA, and continues to operate now as part of DOE, but with its responsibilities broadened to include other than nuclear energy information. TIC provides a centralized facility for information generated by DOE sponsored research. This information is archived, and distributed both within DOE and outside. DOE contractors and the general public share access to the available information according to its classification, which can vary from classified, through limited/sensitive, to unclassified. In addition to domestic sources of information, TIC also scans worldwide energy literature to extract foreign information on energy research, development and application, and to make it available to specialized domestic energy data bases.

Contracts for DOE sponsored research provide, as a rule, report numbers and dates when reports describing the work are due. TIC has a computerized system for keeping track of reports received and for identifying delinquents. Some 125 technical reports are received and processed by TIC every day. They are separated into categories: classified, limited and unlimited distribution, and included in the appropriate data base. Those released for distribution are available in printed form, as microfiche, or as abstracts in a data base searchable on-line. Of the research reports received, some 30 per day are directly distributed according to a predetermined list.(2)

Microfiche copies of DOE and foreign reports obtained through exchange agreements are obtainable from Engineered Systems Inc., P. O. Box 866, Oak Ridge, TN 37830. By photographic reduction up to 84 typed pages of text can be placed on each 14 by 7.4 cm face of one microfiche, so that thick reports can be reproduced on just a few small pieces of transparent plastic.

From 1947 until 1976, TIC was responsible for the preparation and publication of "Nuclear Science Abstracts." Bibliographic records for the entire NSA collection are available for on-line searching as file NSA 2 on the RECON system, and the corresponding abstracts are currently being added to the data base. With the functions of "Nuclear Abstracts" now embedded in the "INIS Atomindex," TIC continues to abstract U.S. nuclear information, and prepares tapes containing appropriate U.S. input to that publication.

TIC is responsible for the preparation and maintenance of the DOE Energy Data Base, described under a separate heading in Chapter 4. The tapes containing that data base are updated twice a month by TIC, and are sent to the National Technical Information Service, at Springfield, VA, for leasing to commercial data base access services ORBIT and DIALOG. Information retrieval from TIC data base collections is possible through an on-line real-time search using the DOE/RECON system.

Input to the various data files constituting the Energy Data Base is obtained through interagency agreements with the Nuclear Regulatory Commission, National Aeronautics and Space Administration, Tennessee Valley Authority and the National Technical Information Service of the U.S. Department of Commerce. Additional input to the Energy Data Base is obtained through contacts with the American Institute of Physics, Science Applications, Inc., IFI Plenum Data Co. and Engineering Information, Inc. Agreements with the International Atomic Energy Agency and the International Energy Agency are supplemented by bilateral agreements with the Federal Republic of Germany, the Netherlands, France, the Nordic countries and Australia for the exchange of technical information.

Whenever DOE contracts result in the production of computer program software, the TIC directs the preparation of appropriate software packages by the National Energy Software Center at Argonne National Laboratory. Such packages include, as a rule, a report describing the method of analysis, a user's manual, a source deck and sample problems. The abstracts of the codes are then added to an appropriate data base, and are made accessible for on-line searching via the DOE/RECON system.

While abstracting the incoming reports for inclusion into the several data bases, TIC publishes hard copy abstract journals. Among publications originating at TIC, and of interest to the nuclear engineering community are:

- *Energy Research Abstracts* (semimonthly abstract journal)
- *Energy Abstracts for Policy Analysis* (monthly abstract journal)
- *Energy and the Environment* (monthly abstract journal)
- *Fusion Energy Update* (monthly abstract journal)
- *Nuclear Fuel Cycle* (semimonthly current awareness bulletin)

- *Nuclear Reactor Safety* (semimonthly current awareness bulletin)
- *Radioactive Waste Management* (semimonthly current awareness bulletin)

The staff of TIC was instrumental in the development of the Federal Energy Data Index (FED or FEDEX) data base for the Energy Information Administration. This data base is maintained by an affiliate of the National Energy Information Center at Albuquerque, NM. It contains information arranged in tables and graphs, rather than text documents.

Textbooks, films and other services, including translations from foreign technical literature, are also among the products of TIC.

For additional information contact: The Technical Information Center, P. O. Box 62, Oak Ridge, TN 37830, tel. (615) 576-1272.

REFERENCES

1. "Technical Information Center, Its Functions and Services," DOE/TIC-4600 (Rev. 3), Technical Information Center, Oak Ridge, TN (May 1982).
2. "Standard Distribution List for Unclassified Scientific and Technical Reports: Instructions and Category Scope Notes," DOE/TIC- 4500 (Rev. 69), Technical Information Center, Oak Ridge, TN (1980).

2.3 NETWORKS USED BY THE NUCLEAR INDUSTRY

Public Domain Communication Networks

Several communication networks have been licensed by the Federal Communications Commission to operate on a nationwide basis. They provide local numbers in all major metropolitan areas for access to data base services located on east or west coast. The networks of interest to the nuclear engineering community are: TYMNET, TELENET, UNINET. The Canadian DATAPAC is also accessible from the United States through interconnection via TELENET. The user dials one of the local TYMNET, TELENET or UNINET numbers to access DIALOG, BRS or the ORBIT system. These networks commonly provide two local numbers each, one for 300, the other for 1200 baud transmission rate (30 or 120 characters per second). All the billing for the use of the communication network lines is done between the data base services and the communication networks without involving the user. The user, however, retains the choice between the two or three communication networks available locally for dialing into the data base service. Conferences are held periodically[1] by the interested parties to assure that both the user

Table 4. Networks with Access to Engineering Data Bases

Network	Address	Telephone
TYMNET	Tymshare, Inc. 20705 Valley Green Dr., Cupertino, CA 95014	(800) 336-0149
TELENET	GTE Information Systems, Inc. East Park Dr., Mount Laurel, NJ 08054	(800) 336-0437
UNINET	Uninet, Inc., Roseland, NJ 07068	(201) 228-8900

and operator interests are taken into consideration in setting up these services.

The major communication networks offering access to a variety of data bases are listed in Table 4. For access of the data bases available in the United States from terminals abroad, most of the networks offer international service. The participating countries are listed in Table 5.

Table 5. Countries Served by the Communication Networks

All Three Networks		TELENET Only	TYMNET Only	UNINET Only
Argentina	Australia			
Austria	Bahrain		Barbados	Barbados
Belgium	Bermuda	Canada	Brazil	
Chile	Denmark	Dominican Rep.	Dubai	Dominican Rep.
Finland	France			Dubai
Hong Kong	Ireland			
Israel	Italy			
Japan	Luxembourg	Kuwait		
Netherlands	New Zealand	Mexico	Mexico	
Norway	Philippines			
Portugal	Singapore			
Spain	Sweden	Saudi Arabia		
Switzerland	Taiwan	U. Arab Emirates		
U. Kingdom	W. Germany			

The NUCLEAR NETWORK

One of the recommendations of the Kemeny Commission, established in the wake of the Three Mile Island accident, was the establishment of an industry-wide international communications network. The nuclear utility industry responded to this recommendation by organizing the NUCLEAR NOTEPAD system. Originally used for communications between the industry and the Institute of Nuclear Power Operations (INPO) in Atlanta, GA, the system has been recently expanded and reorganized.

In January 1984 it was renamed the NUCLEAR NETWORK and its ownership was vested with INPO. It operates similarly to other commercial electronic communication systems, that is with messages originating from local terminals, and the communications traffic routed through a computer supported by Anistics, Inc. of Palo Alto, CA.(2)

The NUCLEAR NETWORK provides direct means of communication between all reactor licensees, major architect-engineer companies, reactor vendors and the principal insitutions set up by the power industry, such as the Edison Electric Institute, the Atomic Industrial Forum and the Nuclear Safety Analysis Center (NSAC) of the Electric Power Research Institute. Access to a dozen of foreign nuclear power organizations is also possible. NSAC is one of the heavy users of the system. Now more than 40 public utility companies are linked this way, and share in transfer of information from and to NSAC. Present level of communication traffic is some 1400 messages per month, with 2 to 3 percent meriting immediate attention.

Local access to NUCLEAR NETWORK is available through the TYMNET system and subscribers are supplied with the local TYMNET phone number, account number and code password. Typical uses of the NUCLEAR NETWORK in day-to-day operation involve transmittal of:

- Upcoming meeting announcements of various task forces, committees, or NRC-sponsored activities.
- Summaries of key points from previous meetings.
- Status of applications for operating licenses and construction permits.
- Significant plant events related to safety.
- Summary of activities related to post–TMI-2 recovery.
- Abstracts of documents added to NSAC data storage system.
- Special interest topics restricted to a select groups of users.

An additional account is available in the NUCLEAR NETWORK system to handle activities related to the Significant Event Program (SEP). This program coordinates contractors, assisting in the evaluation of operating events with impact on reactor safety. Included here are licensee event reports and monthly operating status reports to NRC, as well as the voluntary reports on plant and equipment reliability supplied by utilities to the national data bases. Access to this account is limited to authorized personnel only.

RECON DOE Communications Network

The RECON or REmote CONsole interactive on-line information retrieval system was developed by the Technical Information Center at Oak

Ridge National Laboratory,[(3)–(4)] to facilitate access to data bases developed by the Center. The system is intended for the convenience of the following four categories of institutional users:

- Department of Energy facilities.
- DOE contractors for research, development or demonstration.
- Other government agencies, with missions related to energy use.
- State agencies with responsibilities for statewide energy programs.

The Technical Information Center of Oak Ridge grants access to the DOE/RECON system on approval by Charles E. Spath, assistant manager for information services at TIC. Currently the system has over 700 users. Direct dial-up or access through the TELENET network is available.

RECON is a system for interactive data base access, requiring the user to enter at his console a small number of commands to search the available data bases for information of interest. The typical search will identify sets of documents related by author, subject, year of publication or other criteria. The sets can then be combined using Boolean logic operators AND, OR, NOT to narrow down the material to the specific area of interest. The results of the search can be either displayed or printed out at the user's terminal. A range of retrieval formats is available, varying from brief bibliographic citation only, through full bibliographic citations and abstracts with all descriptors.

Among the data bases accessible through RECON, the following are of particular interest to the nuclear engineering community:

EDB 1	Energy Data Base
NSA 2	Nuclear Science Abstracts (no longer updated)
RIP 3	Research in Progress, Listing
PRD 6	Power Reactor Docket Information
FED 7	Federal Energy Data Index
NSC 8	Nuclear Safety Information
RSI 13	Radiation Shielding Information Data Base
NRC 19	National Referral Center Data Base
RSC 20	Radiation Shielding Computer Codes, Abstracts
NES 23	National Energy Software, Code Abstracts
NTB 29	NASA Tech Briefs
GID 30	Government–Industry Data Exchange Program
EIR 37	Energy Information Resources, Listing
EPD 39	Electric Power Data Base
GEO 40	Geological and Geochemical Aspects of Uranium Deposits
NEA 45	OECD Nuclear Energy Agency Data Bank

Some of these data bases also contain nonbibliographic information, such as table-level records of FED 7, which include descriptions of the data tables contained in the publications cited.

Access to the above and other data bases, for users not qualified for RECON membership, is possible through commercial DIALOG, BRS and ORBIT retrieval services. The inclusion of a particular data base among those offered by the commercial data base access services is dependent on their assessment of public demand plus commercial viability, and may vary from time to time. Recently, for instance, BRS discontinued offering the Energy Data Base for access.

Livermore Magnetic Fusion Computing Network

Among network developments already in place is the Magnetic Fusion Energy Computer Center (MFECC) and its associated computing network, operating at Lawrence Livermore National Laboratory under the direction of John Killeen.

The Center was established to extend the analytical capabilities of the fusion community in the United States by concentrating in one place the best computer hardware and software currently available. The Center and the network are equipped with the most up-to-date computing equipment, of capacity and speed as yet unavailable at many other centers for fusion research and development. The equipment of the center comprises two CRAY-1 machines and one CDC-7600 machine. The file storage capacity consists of 2400 megabytes of disk space shared between the CRAY machines and the CDC-7600. In addition, MFECC operates the CDC-38500 mass storage device which has 8000 storage cartridges capable of holding 500 billion bits of information. The last line of storage consists of a 1600 bpi magnetic tape library. Remote User Service Stations (RUSS) can access MFECC through TYMNET lines, in 300 or 1200 baud transmission mode, and interface with the mainframes through PDP11/44 hardware.

Access to the MFECC computing facilities, computer codes deposited with the Center and the associated nuclear property data bases of the Livermore Magnetic Fusion Data Center is available for qualified users through the network of links between the Center and the major users. Access qualification for the users requires that they be engaged in the fusion program, either at a national facility or under the sponsorship of one of the DOE contractors. The security of access lies in the clear definition of the network of centers involved in the fusion program. Among the users there are 27 universities, 11 national laboratories and 8 major industrial contractors listed in Table 6.

The capabilities of the computing network allow the user several options.

Table 6. Users of the Magnetic Fusion Energy Computer Center

AMAF Industries, Inc.	Univ. of Illinois	New York Univ.
Argonne Nat. Lab.	Univ. of Iowa	Northwestern Univ.
Univ. of Arizona	Jaycor	Oak Ridge Nat. Lab.
Auburn Univ.	Johns Hopkins Univ.	Penn. State Univ.
Brookhaven Nat. Lab.	Lawrence Berkeley Lab.	Princeton P. P. Lab.
Calif. Inst. of Techn.	Lawrence Livermore Lab.	Rensselaer Pol.Inst.
U. of C., Berkeley	La Jolla Inst.	Science Appl. Inc.
U. of C., Irvine	Los Alamos Nat. Lab.	Stanford Univ.
U. of C., Los Angeles	Univ. of Maryland	TRW
Columbia Univ.	Mass. Inst. of Tech.	Univ. of Texas
Cornell Univ.	Math. Sciences NW	U.S. Dept. of Energy
Dartmouth College	McDonell-Douglas Co.	Univ. of Washington
General Atomic Co.	Univ. of Michigan	Westinghouse El. Co.
Georgia Inst. of Tech.	Univ. of Missouri	C of William & Mary
Hanford Eng. Dev. Lab.	Nat. Bur. Standards	Univ. of Wisconsin
Idaho Nat. Eng. Lab.	Naval Res. Lab.	Yale Univ.

He can upload the input files for his calculations from a computer terminal console located at his laboratory, and execute them on the Center's mainframe computers using one of the computer codes held by the Center in its library. Alternately, he can mail beforehand a magnetic tape bearing his own program, and request that it be mounted on the system prior to uploading the input data. In principle, the program itself could also be uploaded, i.e. transmitted from the local installation to the mainframe, but except for short programs, this involves longer transmission time than most people are willing to devote to this operation. Once the calculations are done, the numerical or graphical output can be downloaded from the mainframe to the user's terminal for viewing, or for printing out at the local line printer. For a voluminous output, hard copy or microfiche can be requested by mail. The operation of above system has been very successful, with mainframe utilization factors exceeding 90%.

It may be noted here that Lawrence Livermore National Laboratory has been instrumental in other networking initiatives. In particular, its local data system network was taken as a model for the National Cooperative Property Data Network, which is now being set up.

The MFECC publishes a monthly bulletin, *Buffer* and distributes it to the users of the system on its mailing list.

The address of the Center is: Magnetic Fusion Energy Computer Center, University of California, Lawrence Livermore National Laboratory, Livermore, CA 94550, tel. (415) 422-4026.

REFERENCES

1. *Proc.* of the Third Berkeley Workshop on Distributed Data Management and Computer Networks, Lawrence Berkeley Lab. (Aug. 29–31, 1978).
2. Paul Vlajcic, "Building a Network of Safety," *Nuclear News* 27, 76–81 (Mar. 1984).
3. "ENERGYINFO/ON-LINE, The DOE/RECON System," DOE/TIC-4609 (Rev. 1), Technical Information Center, Oak Ridge, TN (May 1982).
4. J. Robinson, C. Backhus, R. LaBrie, and L. Spurrier, "DOE/RECON User's Manual Data Base Summaries," LBL-12744, DOE/TIC-4586/R1, Part 3, Summaries, Lawrence Berkeley Lab. (July 1982).

2.4 LEGAL AND BUSINESS ASPECTS OF CODES AND DATA

The early development of computerized analytical methods of nuclear engineering arose from federally sponsored projects. When not related to defense matters, the resulting software and methods remained in the public domain, and were accessible without fee or purchase agreements to qualified, bona fide users. With the rise of the commercial sector of the industry, the software developed for reactor analysis became a salable commodity. The commercialization of mature reactor concepts brought with it proprietary computer codes, tailored, as a rule, to fit the specifics of the reactor hardware and offered for sale to reactor owners.

The developmental nature of much of the work accomplished during the past decades within the domain of nuclear engineering carried with it the need for continuous updating, modification and improvement of analytical methods and for sharing the achievements of one group with others active in this field. In the early stages, the application of copyrights and service fees was pointless and counterproductive to growth. Once the developmental stage was passed, however, proprietary considerations appeared. Yet, even at this late stage, it is fair to say that rare is the code that has not undergone numerous changes and improvements, either directly by users or at users' suggestion. In the process of code modification the questions of authorship, ownership of the improvements and the propriety of transmitting these improvements to other users have become inextricably intertwined.

Some precedents have already been established with regard to responsibility for the content of reports conveying the results of research or code development. It has become the rule to include on the back of the title page a disclaimer in a standard form stating that the authors bear no responsibility for the accuracy of the content or for the consequences of its use. When the material contained in such a report is transferred to a magnetic medium either as a computer code or for inclusion in a data base, the disclaimer may be assumed to transfer to the magnetic medium as well, even though it rarely appears as a part of it.

The precedents that have evolved with regard to the ownership of nuclear engineering software connected with defense but available for distribution include restrictions on users. A typical example concerns the use of neutron diffusion theory code PDQ. This is one of the many diffusion theory codes developed for use in the design of reactor cores. Most such codes are in the public domain, and indeed many foreign equivalents are in existence. Nevertheless, because of the origin of PDQ in the naval reactor program, and its continuing use in this capacity, its distribution abroad has been restricted. A certain percentage of other computer codes fall into the same category.

Many of the above considerations apply to data bases as well. Intimately connected with the concept of data bases are questions of ownership, right of access and legal responsibility for the quality of their contents. So far the questions of ownership of data bases have not been treated uniformly. The uncertainty centers on whether the body collecting, collating, formatting and possibly evaluating the data from outside sources acquires thereby a right to its ownership. Because usage fees are charged and wholesale copying is resisted, the status of ownership is recognized, but its legal implications have not been fully tested. Thus far, the royalties charged the users are treated as recovery of the cost incurred in assembling the information and putting it in usable format while disclaiming responsibility for the content. As long as the content is bibliographic in nature, this approach is acceptable.

As more data bases are beginning to include numerical information in the form of tables of property data, performance records or test results, the question of evaluation of the quality of data comes to the fore. This is particularly important with regard to the the proposed National Cooperative Materials Property Data Network. The present technical means for computerization of existing collections need to be matched by an effort to evaluate the data at a level adequate to ensure the quality of the data base content.

Copyrightable Materials

As mentioned above, many questions of ownership, and thus copyright assignment, arise in connection with the use of data bases and computer codes. It is now recognized that computer software developed by private means does constitute a copyrightable material.(1) It must be stressed here that the copyright law(2) does not protect the ideas or the analytic equations embodied in the code, but it does protect the exact form of expression of these ideas in the coded material. Within the intent of the law, license fees may be charged by the developer for software use, prorated on a time basis, or as one-time payment, and warranties may be imposed on the user to prevent him from copying and distributing the software beyond his own usage. The extent to which these restrictions can be enforced is currently being tested in the courts.

Legal tests of ownership right require the ability to compare and identify complex coding material, possibly altered by a change in its title, appearance and even arrangement, to conceal the fact that the material has been copied from a proprietary source. Many such tests of ownership have been brought to court in the past decade, and a substantial body of adjudication has accumulated to serve as a guide to owners and developers of software.

Legal services have sprung up to assist the interested parties, and a specialized publication, *Computer Law Service,*[3] appeared in 1972. Since that year seven volumes of the *Reporter* part of this publication and five volumes of *Articles* dealing with background information have been issued. The publication summarizes past cases of litigation and reprints court decisions, as well as the arguments of counsel for the plaintiff and the defendant.

Formally the publication is divided into parts covering the following areas:

1. The Law and the Computer (federal and state laws given verbatim)
2. Legal Problems of Systems Design
3. Computer System Procurement
4. Protection of Proprietary Rights
5. Liability Aspects of Computer Usage
6. Data Communications
7. Other Governmental Aspects
8. Legal Aspects of Particular Industrial Applications
9. International Aspects

Over the past decade the collection of loose-leaf issues of this publication has grown to cover the whole gamut of litigation in matters pertaining to development, distribution and sale of computer equipment and services. A good part of this information relates to the development and marketing of computer software, and thus is of interest to people in the nuclear industry who are responsible for maintenance and improvement of computer software obtained from outside sources, and for the sale to outside users of the software developed in-house.

Software and Information Access Fees

In times when the development of nuclear power was carried out under the auspices of the federal government only, the services of collecting, duplicating and distributing computer codes were centered in the national laboratories and were offered free of charge to federal agencies and contractors. Following the entry of commercial companies into the field, and the establishment of the National Energy Software Center at Argonne, IL, small fees started to be charged for the services rendered. In 1982 an annual subscription fee for regular users and full recovery of costs were imposed.

If, in addition to software, the data base use fees are included in the financial arrangements of information transfer, three categories of software and user fees can be distinguished at the present time:

1. The reproduction and service fees charged by federally sponsored information centers and computer code repositories.
2. Service fees charged to the users of on-line data banks and software exchanges.
3. User-supported software production through voluntary contributions.

The first category listed above is well represented by fee arrangements introduced in July of 1982 by the National Energy Software Center (NESC) at the behest of the Department of Energy. It consists of a yearly subscription charge, covering the cost of the first two library packages supplied by NESC to the participating installation, the code catalogs and their periodic updates. In addition, there is a per-package charge for succeeding library packages requested from NESC within any given subscription year. For the purpose of administering the system of charges introduced by NESC, the user installations were divided into three classifications of the yearly subscription fee:

1. DOE or NRC offices and their cost-type contractors: $1350 fee.
2. Other federal agencies; educational institutions; nonprofit organizations; local, state and foreign governmental bodies: $1458 fee.
3. Commercial entities: $1785 fee.

These basic fees cover the period from August 1, 1982 to September 30, 1983. The fees for the third and succeeding code packages are also arranged in the same three classes and vary, depending on the length of code requested, from $135 to $1395 in class 1, from $145 to $1506 in class 2, and from $178 to $1844 in class 3. In cases when the requestor wishes the information to be transmitted on magnetic tape, a blank tape must accompany his order. As a rule, the codes are supplied to the user in the form of a source listing in FORTRAN, plus sample problems, although versions already compiled for a specific machine can also be requested. A price list for all the codes handled by NESC will be supplied on request.

The second category of user fees is charged for access to data bases. The introduction of these fees is justified by the need to recover the operating cost of such a data base. In commercial ventures of this kind access fees are charged in almost every case, the only exception being the various bulletin boards or similar service arrangements. Such considerations apply to a lesser degree in the case of publicly funded data bases, although even there the desire to lessen the financial burden of the sponsoring organization dictates fees.

The use of data bases offered through commercial arrangements may involve several alternatives. The data base itself may be created, owned and operated by an organization that does not want to relinquish control over it, but merely makes it accessible to user services for a fee. In the case of data base access achieved by the user through the public long distance telephone system, the user pays the data base usage fee to the owner, and the telephone charges are billed directly to the user by the telephone company.

When the data base owner does not want to handle billing, the data base tapes may be rented to one or more of the information services. If such is the case, and local access to the rented data base is made through a private telecommunication network, the information service is the agency collecting the usage, telecommunication and royalty fees. The information service, in turn, pays a royalty to the data base owner and usage fees to the telecommunication networks, and retains the data base access fee. In this arrangement the same data base may be available through more than one information service, each of them in turn offering it to its subscribers and using the same communication networks. In some cases the information services create their own data bases, thereby saving the user the royalty component of data base connect costs. This, however, is not the most common arrangement.

Considerable differences exist in user charges, depending on the type of data base searched, length of connect time and information transmission rate. Different charges apply when use is made of dedicated or public communication lines. When access is made through commercial information services, some sort of an initial sign-on fee is required, which may be as little as $50, followed by periodic billings for connection charges and network use fees.

The average cost of a typical search of a data base is now estimated to be about $32, but this user fee figure conceals the fact that the hourly connect rates may vary, depending on the data base, by more than a factor of ten. For example, connection fees to the DIALOG network vary between $25 and $300 per hour. To avoid excessive costs of searching incurred through lack of skill in performing the search, most information services offer tutorial sessions at rates of $20 to $30 per hour, plus access to a test file on which the various data base search techniques can be practiced by the beginner. Addition of "after hours" service, offered at reduced rates by DIALOG and BRS information services, reduces the above costs considerably for noninstitutional users willing to do their data base searches in the evening or at night.

The category of voluntary contributions for user-supported software emerged only recently, and is based on the so-called freeware concept. In this category, the software, primarily intended for microcomputers and intelligent terminals, is supplied at cost of the medium on which it is being transferred, generally a floppy disk. Encoded internally is a message that identifies the author of the software and his address, and suggests the size of a voluntary authorship contribution, which should be sent directly by the user to the author after full satisfaction with the operation of the software is obtained. Further reproduction and dissemination of the software is suggested, but modifications to it are discouraged. This is in essence an attempt to return to an earlier time in the computer era when programs were still obtainable for the asking, and when the cost of producing them was relatively modest.

The freeware concept of user-supported software development was introduced with the PC-TALK code for computerized asynchronous telecommunications, written by Andrew Fluegelman and distributed by "Freeware," The Headlands Press Inc., P. O. Box 862, Tiburon, CA 94920. This program handles on-line searches of data bases from intelligent terminals, an operation of interest to professionals in the field of nuclear engineering who wish to access data bases for information on reactor design and operation.

Data Base Security

Considerations of data base security enter into play at several levels. Most basic is the question of data integrity, and protection from inadvertent damage. On a higher level, access to and modifications of the data base content may be restricted to individuals authorized to perform the updating and purging tasks, thus eliminating any possible duplications of effort and conflicts of authority.

In large installations housing and operating data bases, the tasks of entry, update and purging of information follow highly structured procedures, subject to cross checks and administrative controls. In installations dealing with confidential or proprietary information, these administrative operating checks may be supplemented with requirements for personnel clearances, in either the formal or the background reference sense. In addition to these cross checks at the human level, there are mechanized software or hardware security measures attached to the operation of practically every data base. It will, therefore, be proper to mention them here in outline.

For communications with data bases over public telephone lines three levels of data base access security can be distinguished. The most basic level consists of assigning each user a confidential password and budget account number that must be entered into the system at each interaction before the response of the system to the user's query can be elicited. This approach is used by most commercial data base access services. The assigned password and budget number can be used from any entry point, be it a dedicated terminal or a public telephone used in connection with a portable modem. Apart from the confidential password and account number, the rest of the transaction between the user and the data base is carried out in ordinary ASCII syntax. To assure the security of the password and the budget number, many of the terminal systems obliterate the typed-in characters from display the moment they are entered into the system, thus avoiding inadvertent disclosure.

At the next level of security, access to the data base is limited to password holders calling from specific telephone numbers. This is termed a call-back system.[(4)] In this system, the computer handling data base access responds to the user's password in the incoming call by disconnecting the line. It then consults a list of matching passwords and phone numbers with which it is authorized to do business. If the incoming password is found, it extracts the authorized phone number, and calls back with a response. The list of passwords and the matching phone numbers is physically contained in a Secure Access Unit, located outside the computer, and can only be programmed manually by the authorized data base operator. An audit-trail is kept of the successful and rejected connects. The security of this system depends, therefore, on limiting access to the authorized terminals and not listing the corresponding phone numbers in the public telephone book.

The highest level of security still obtainable with the use of public telephone access is possible with data encryption. Various commercially available encryption systems, not requiring specialized hardware, have been described in the open literature and implemented in communication networks requiring some degree of confidentiality. A popular system for the protection of commercial computerized data systems is the TOP SECRET computer program[(5),(6)] developed by CGA Products Group of 255 Route 520 East, Marlboro, NJ 07746. This second generation state-of-the-art security package requires no system modification, and is very user-friendly. It is mechanized on an IBM 370 machine.

The two best known algorithms for implementing a public-key cryptosystem are the "knapsack" system of Diffie and Hellman[7] and the RSA algorithm developed by Rivest, Shamir and Adleman.[8] Both encrypting systems use certain well-known mathematical problems for which solutions are possible only if certain "trapdoor" parameters are known. Otherwise, decryption is very time-consuming, if not impossible. The security of both systems depends on the excessive time required to solve the implicit problem without the knowledge of the "trapdoor" parameter. Breaking of such an encoding-decoding system requires a considerable degree of mathematical sophistication and computing capability, usually outside the realm of the resources available to an individual bent on creating mischief.

Perhaps the highest level of encryption available to the public is represented by the Data Encryption Standard, produced by the National Bureau of Standards. It achieves encryption by scrambling the transmitted message through iterative multiplication with a 56-bit key. The key is randomly generated and has 8 data bits added to it for parity checking, thus becoming 64 bits long. The keys are highly secret and must be guarded carefully. Implementation of the system requires proper circuitry at both ends of a link, thus limiting the use of the system to dedicated installations.

REFERENCES

1. *Computer Programs and Data Bases: Protecting and Licensing Proprietary Rights After the 1980 Copyright Amendments*, Proc. of a Conference on Law and Business, Daniel T. Brooks and Michael S. Keplinger co-chairmen, Harcourt Brace Jovanovich, New York (1981).
2. Title 17, United States Code, U. S. Copyright Law. Citations 502, 504, 602 and 603.
3. *Computer Law Service*, Calaghan and Company, Chicago, Il. *Reporter*, Vols. 1–7 (1972–1980); *Articles*, Vols. 1–5, (1972–present).
4. C. Gillard and J. Smith, "Computer Crime: A Growing Threat," *Byte*, 398–424, (Oct. 1983).
5. J. B. Goldberg, "TOP SECRET Helps Bank Centralize Administration," *Bank Systems and Equipment*, 20, No. 6, 67–68 (June 1983).
6. "TOP SECRET," CGA Software Products Group, Marlboro, NJ (1981).
7. Whitfield Diffie and M. Hellman, "New Directions in Cryptography," *IEEE Trans. PGIT*, 22, No. 6, 644–654 (Nov. 1976).
8. R. L. Rivest, A. Shamir and L. Adleman, "A Method for Obtaining Digital Signatures and Public-Key Cryptosystems," *Communications of ACM*, 21, No. 2, 120–126 (Feb. 1978).

3. DATA CENTERS OF NUCLEAR ENGINEERING

3.1 NUCLEAR DATA CENTERS AND THEIR HOLDINGS

Collections of specialized information have existed in the United States for many years, usually in the form of libraries attached to universities or in scientific and engineering societies. The National Bureau of Standards and the Patent Office have long represented the federal government in this field. The creation of national laboratories, in response to the exigencies of defense during the World War II, added an important component to this picture, while shifting the emphasis from collections of books and articles to collections of experimental data and research reports. This shift in emphasis led, in turn, to the use of new terminology. The designation "data base," now frequently written as a joined word, "database," was created to describe the new collections.

In broad terms, the present day data bases consist of collections of documents, journal articles and their abstracts, and bibliographic compilations, as well as collections of numerical data, often in tabular form. In selected instances, graphic materials, be they drawings or photographs, are also included. Of interest here are the data bases of science and technology, that contain information pertinent to nuclear engineering. Capitalizing on the heavy use of computer technology in nuclear engineering, many of these data bases are in the process of transition from a static repository of printed or typed information to a dynamic form of information stored on a magnetic medium, where it can be handled, searched and transmitted electronically.

Once the information is entered on a magnetic medium in proper format and equipped with relevant keywords, or numerical code words, it can be loaded into a computerized retrieval system. The data files thus created can be queried either off-line, by data base personnel, or on-line, by the end user. Those data bases that provide an on-line searching capability help extend greatly the resources readily accessible to the nuclear engineer, provided the engineer is equipped with a suitable computer terminal, connected to an appropriate communications network. Some of the data bases with on-line searching capabilities are available at national laboratories, and some are accessible through commercial networks.

Recently, the engineering societies and technical publishers have entered the field of data base handling. Starting with service offering access to computerized bibliographies, some data bases and networks have expanded to include an abstract search feature, whereas some have included the computerization of the full text of resource materials.

Early History of Unclassified Nuclear Data

The early measurements of neutron and photon cross sections were made at various university laboratories, both at home and abroad. Studies of photon cross sections and compilations of their values were made in connection with early X-ray work. The National Bureau of Standards served as a convenient reference point, together with the International Commission on Radiological Protection (ICRP). Initially, there was no corresponding collecting body for neutron cross sections, in part owing to the scarcity of neutron sources in the era before the advent of fission reactors.

In the wake of the Manhattan project, national and industrial laboratories undertook to measure the needed neutron cross sections, and selected sites were designated as repositories of nuclear cross sections. Several data centers were set up. The specific tasks assigned to each data center were those of collecting, evaluating and disseminating the information in its custody.

The cross-section data deposited with the centers were formerly hand-processed, evaluated on site, tabulated and plotted. They were then distributed in the form of reports. Of the above tasks, the evaluation of data has proved to be the most challenging to their staffs, because of the ever increasing flood of information entering the centers, and the difficulty of mechanizing this phase of the work.

To ease the task of data acquisition and evaluation, efforts were made to enlist the participation of users in various phases of these operations. The success of the resulting cooperative efforts in the production of evaluated cross sections and collapsed cross-section sets has been most gratifying. The details of these cooperative arrangements will be described below. The entire effort could very well be considered as a model for other fields of data collection and evaluation, that are badly in need of consolidation but still await the organizing initiative of measurers and users. One such field is thermodynamics, with its fragmented national data base.

The major effort mounted in the past three decades to provide a data base for the development of the nuclear industry has resulted in the creation of valuable data collections of nuclear and material properties, and has given rise, as well, to the field of equipment performance assessment, risk evaluation and environmental impact studies. All these disciplines have enriched the engineering sciences and enlarged the U. S. data base.

The following sections describe the major national data repositories of nuclear and material properties, as well as computer software for data handling and nuclear design.

Brookhaven National Nuclear Data Center

The Brookhaven National Nuclear Data Center, located on Long Island at Upton, NY 11973, holds the distinction of being the pioneer in the collection of and distribution of neutron cross section data. It was originally known as the "Sigma Center." With time, the activities of the Center grew in importance and gained international recognition. A Cross Section Evaluation Center was initially operated as a separate entity, but on September 15, 1967 the two centers were merged to form the National Neutron Cross Section Center, which was subsequently turned into the National Nuclear Data Center (NNDC).

The earliest workers in this field, associated with Sigma Center and the release of unclassified nuclear cross section data, were Donald J. Hughes and John A. Harvey. Their names have passed into history, and the first edition of their report, "Neutron Cross Sections,"[(1)] BNL-325, issued by the Brookhaven Laboratory in 1955, in 10″ by 15 1/2″ format, is now a collector's item. Hughes published at the same time a companion monograph of identical title,[(2)] to explain much of the technique and rationale used in developing the content of this early data base. Succeeding editions of this report are being produced to this day under the same title. They appear, however, in multivolume form and in a different format. The work of Hughes and Harvey was subsequently continued by the National Nuclear Data Center staff.

From the initial distribution of printed reports containing hand-generated cross-section plots, the activities of the Center expanded into providing qualified users with nuclear data, also in a form compatible with computer use. The development of computer codes for data evaluation and checking followed. The activities dealing with data evaluation and checking were expanded and the cooperation of the users was enlisted in a meaningful and significant way. Meanwhile, the graphic reproduction of data became mechanized, and cross-section plots and tabulations produced by computer became the rule, appearing in the successive editions of the two reports "Neutron Cross Sections," [(1)] BNL-325, and "Neutron Cross Sections, Angular Distributions," [(3)] BNL-400. For the most part, the computer software required for the extensive data handling and plotting was developed in-house, greatly simplifying the production of reports and improving their graphical form and quality to the point where they have now become models in the nuclear field.

In 1965 an effort was initiated to produce a consistent, evaluated data set for nuclear applications. This was the origin of the Evaluated Nuclear Data File (ENDF), actually a collection of data files in a computer-retrievable form. Because of the magnitude of the task, it was thought that a joint effort with the users' groups was required. This led to the formation of CSEWG, the Cross Section Evaluation Working Group, chaired by S. Pearlstein. The group specified formats for dissemination of cross-section data, assigned evaluation tasks to the various participating groups from industry and national laboratories, and coordinated their efforts at regular semiannual meetings. The product of their efforts, ENDF/B-I, the Evaluated Nuclear Data File of Brookhaven, made its long-awaited appearance in 1967. This data file is now divided into parts A and B, and currently appears in version V. The specifications for ENDF/A are described in the report BNL-8381,[4] those for ENDF/B in BNL-50066,[5] both authored by Henry C. Honeck. Through the process of introducing successive versions of ENDF, the format specifications of ENDF/A and ENDF/B files are now the same. The contents of the files reside on magnetic tapes in custody of NNDC. Table 1 below gives the year of appearance of a given version; Table 2 compares the characteristics of parts A and B; Table 3 lists the quantities contained in files of ENDF/B.

For the purpose of data retrieval, ENDF/B is divided into data files identified by the MF number. Numbers 1–10 refer to neutron reactions, numbers 12–27 to photon reactions, as shown in Table 3. Not all MF files are currently used, as indicated in the table.[5]–[8] In addition to cross-

Table 1. Chronology and Characteristics of ENDF Tapes

ENDF/B	Characteristics of Version	Released
I	Existing evaluations converted to ENDF format	1967
II	Modified after comparisons with differential and integral data	1970
III	Improved higher energy data for shielding	1972
IV	Uniform coverage from 10^{-5} to 2×10^{7} eV plus decay data for 800 nuclides added	1974
V	Covariance files, gas production file and activation data files added	1979
V/Mod II	Selected ENDF/B-V data revised	1983

Table 2. Comparison of Characteristics of ENDF/A and /B Files

Characteristics	ENDF/A	ENDF/B
Basic storage unit	Evaluated point cross section data covering a particular energy range for one reaction type and one material. Format the same as ENDF/B in most cases, or as received from authors.	All evaluated point data for one material needed for a reactor calculation. Simple format.
Type of data included	All reaction types for incident and final particle types.	Data for neutron-induced reactions required for reactor calculations.
Ordering of data	Data stored in the order in which received by the ENDF center.	Ordered by material no., data type and reaction type.
Selection and revision of data	No selection made. All data accepted and added to master files. Contains many alternate evaluations.	One complete set of data for a material selected and stored; updated at regular intervals.
Main usage	Storage of partial evaluations used as input for generation of complete evaluations.	Copied and distributed to users for input to reactor codes or used to generate multigroup, multitable cross section sets.

section data, some files are set aside for storing the covariances, generated in a joint effort with the radiation shielding group at Oak Ridge for cross sections, neutron yields, resonance parameters, and so on.

An example of cross-section data plots obtained from the data contained in the ENDF/B file is given in Figure 4.

At present, the neutron cross-section data are handled under two headings: CSISRS Library, containing experimental data, and ENDF Library, containing evaluated data. Together, these collections contain well over 2,000,000 cross section points and 150,000 resonance parameters. Major reactor designers and producers, as well as the national laboratories have developed an impressive array of subsidiary computer programs. They handle the data contained in the ENDF/B, and perform various flux-averaging and group collapsing operations needed to produce cross section libraries in formats suitable for use in locally developed design meth-

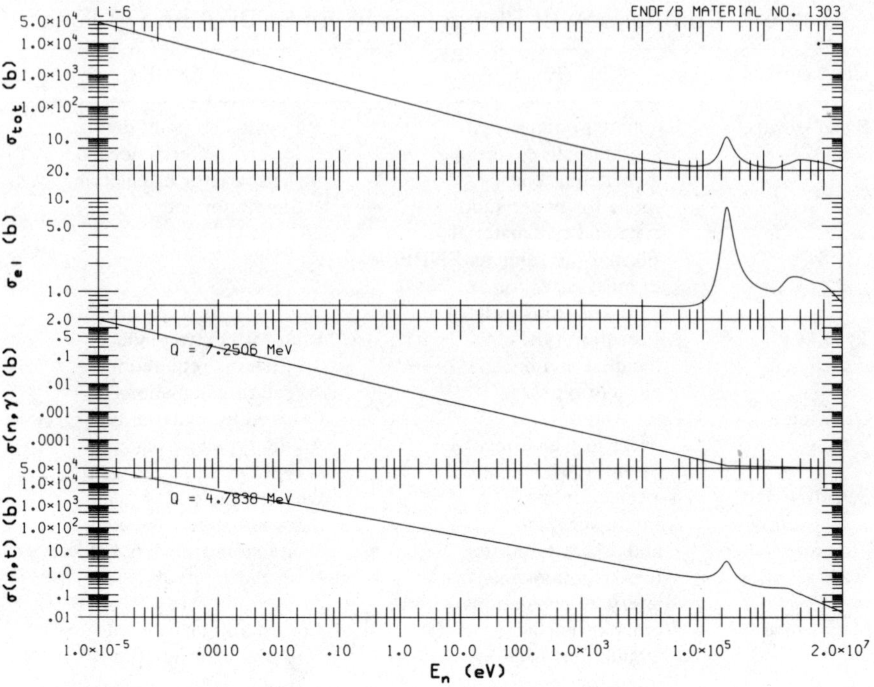
Li-6
ENDF/B MATERIAL NO. 1303
σ_{tot} (b)
σ_{el} (b)
σ(n,γ) (b)
Q = 7.2506 MeV
σ(n,t) (b)
Q = 4.7838 MeV
E_n (eV)

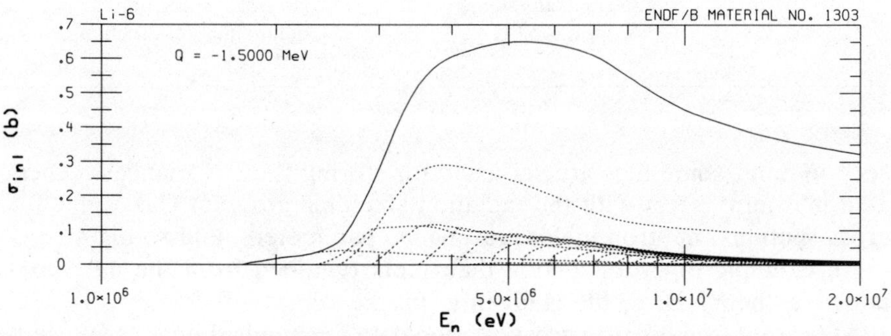
Li-6
ENDF/B MATERIAL NO. 1303
σ_{inl} (b)
Q = -1.5000 MeV
E_n (eV)

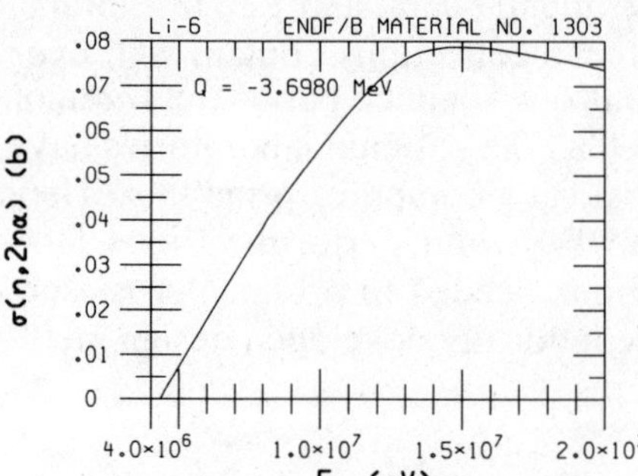
Li-6
ENDF/B MATERIAL NO. 1303
σ(n,2nα) (b)
Q = -3.6980 MeV
E_n (eV)

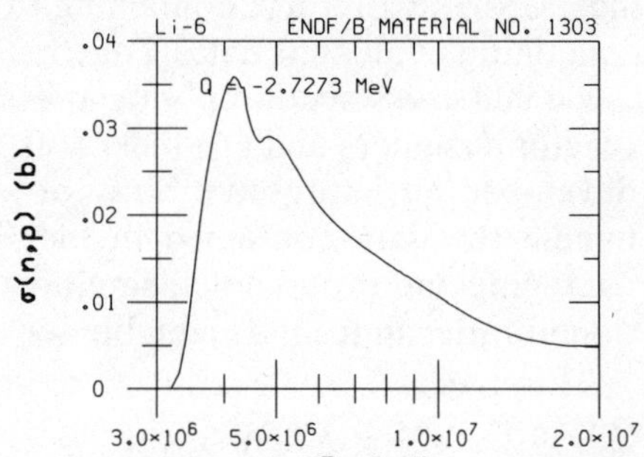
Li-6
ENDF/B MATERIAL NO. 1303
σ(n,p) (b)
Q = -2.7273 MeV
E_n (eV)

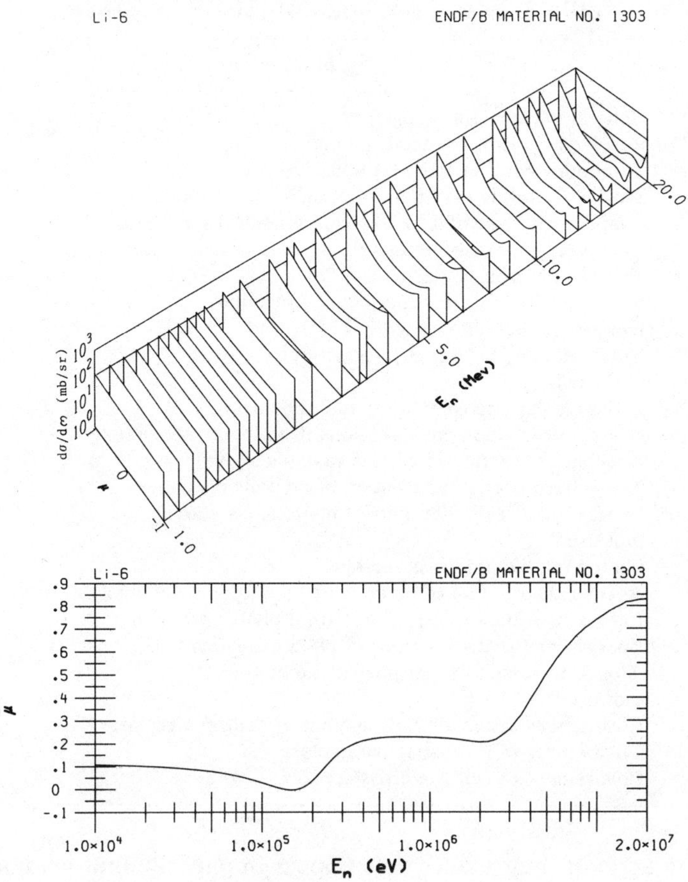

Figure 4. An example of ENDF/B cross-section plots.

ods. The titles and functions of the principal codes are listed in Tables 4 and 5. Such variety of data handling and formatting codes for cross sections is indicative of the variety and lack of standardization in this field. Many of the major cross-section users have introduced their own formats, group energy breakpoint schemes and code library arrangements. Foreign developments added further complexity. Some effort at standardization of energy group structure of the various cross-section sets has been initiated to remedy this situation.

NNDC publishes the *Nuclear Data Sheets*.[9] They contain nuclear level schemes, energies and intensities of emitted radiation, and generic relationships between the members of radioactive decay chains. The

Table 3. Files Contained on ENDF/B-V Tape

MF No.	File Content
1	General information, integral data
2	Resolved neutron resonance parameters
3	"Smooth" neutron cross sections
4	Elastic scattering distributions, Legendre polynomials
5	Evaporation spectrum for secondary neutrons
6	Energy-angular distributions for secondary neutrons
7	Thermal neutron scattering law data
8	Radioactivity from neutron-induced reaction
9	Isomer production probability
10	Cross sections for isomer production
11	(not used)
12	Radiative capture gamma ray multiplicities
13	Gamma production cross sections (from neutron reactions)
14	Angular distribution of gamma rays (as above)
15	Normalized energy distribution of photons (as above)
16	Energy-angular distributions of photons (as above)
17–22	(not used)
23	Photon interaction cross sections
24	Angular distribution of photons (from photon reactions)
25	Energy distribution of photons (from photon reactions)
26	Energy-angular distributions of photons (as above)
27	Atomic form factors (for photon interactions)
28–30	(not used)
31	Covariances of the average number of neutrons per fission
32	Covariances of resonance parameters
33	Covariances of neutron cross sections

sheets, in spite of their name, now appear in paper-bound volumes, with Volumes 38–40 published in 1983. Their editor is M. J. Martin.

The national fusion program can be credited with starting an effort to standardize the format of cross-section libraries used by engineers and designers in this field. A similar standardization effort in the fusion field attempts to prevail upon contractors and designers to use SI units in their work by specifying this requirement in contracts let by the Department of Energy.

REFERENCES

1. Donald J. Hughes and John A. Harvey, "Neutron Cross Sections," BNL-325 (1st edit. 1955, 2nd edit. 1958 with supplements in 1960 and 1966, 3rd edit. 1976 with tabulation of neutron resonances in Vol. I, and cross-section curves in Vol. II).
2. Donald J. Hughes, *Neutron Cross Sections*, Pergamon Press, New York (1957). (Descriptive companion to the BNL-325 report.)

Table 4. Processing Codes for ENDF/B-IV and ENDF/B-V Libraries

Code	Type of Data in Library and Code Serviced	Origin
AMPX (modular)	Neutron interaction, photon production and interactions for ANISN, DOT, MORSE, KENO codes	ORNL
BDASS (modular)	Neutron interaction, photon production and interaction, fission product for the JOSHUA system	SRL
ENDRUN	Neutron interactions and shielding factor tables for TDOWN code	GE
ENDRAFL	Epithermal neutrons for RAFFLE Monte Carlo code	EG&G
ETOE-2	Multigroup library for MC^2 code and fine group cross sections for SDX code	ANL
ETOG-5	MUFT, GAM, ANISN, LASER codes	WNES
ETOP-15	Neutron interactions for GAM code	EG&G
ETOMX	Neutron interactions for MUFT and GAM codes	BAPL
ETOT	Thermal neutron cross sections, GE lattice codes	GE
ETOX	Neutron interactions for IDX code	HEDL
FLANGE-II	Coherent and incoherent thermal scattering kernel (works only for the ENDF/B-III version)	SRL
GFE4/GAND3	Neutron interactions, GGC-4, GGC-5, MICROX codes	GA
MACK	Kerma factors, activation, photon production for MACKLIB	ANL
MINX	Neutron interactions and shielding factors for SPHINX code, versions for IBM and CDC machines in CCCC, AMPX and MATXS formats	LASL ORNL
NJOY	Fast and thermal neutron interaction, photon production, photon interaction, heat production and covariance data in CCCC, DTF, MATXS formats	LASL
PUFF	Covariance data for the FORSS system	ORNL
RESEND	Infinitely dilute, 0^0K pointwise cross sections for SIGMA1 and INTEND codes	BNL
SAMX	Neutron cross sections for SAMF continuous energy Monte Carlo code	MAGI
SUPERTOG	Epithermal neutron interactions for ANISN, DOT and MORSE codes	GE
UNCER	Covariance data for CTR applications	UW
VIM codes	Neutron interactions for VIM continuous energy Monte Carlo code	ANL

3. Donald J. Hughes and Robert S. Carter, "Neutron Cross Sections, Angular Distributions," BNL-400 (1st edit. 1956, 2nd edit. 1962, 3rd edit. 1970).
4. Henry C. Honeck, "ENDF–Evaluated Nuclear Data File, Descriptions and Specifications," BNL-8381 (June 1964).
5. Henry C. Honeck, "ENDF/B–Specifications for an Evaluated Nuclear Data File for Reactor Applications," BNL-50066 (ENDF-102), (May 1966, rev. July 1967).
6. M. K. Drake, rev., "Data Formats and Procedures for the Evaluated Nuclear Data File Library, ENDF," BNL-NCS-50496 (ENDF 102) (2nd edit. (ENDF/B-V), Oct. 1979).

Table 5. Codes Maintained by NNDC to May 13, 1983

Code	Function	Origin
CHECKR	Checks ENDF data files for legal formats	NESC
CRECT	Edits card input records for ENDF data files	NESC
FITZCON	Checks ENDF data files for physics consistency and adherence to recommended practices	NESC
INTER	Calculates resonance integrals and spectrum averaged cross sections from ENDF data file	BNL
LISTFCV	Produces an annotated listing of ENDF data file	BNL
PLOTFB	Lists and plots ENDF data	NESC
PSYCHE	Performs more complex physics analyses of ENDF data files	BNL
RIGEL	Converts between different ENDF storage records and merges data files	NESC
STNDRD	Converts ENDF formatted data files to standard data representation and generates a dictionary	BNL
SUMRIZ	Creates an abstract summary of the contents of an ENDF data file	BNL

7. R. Kinsey, comp. : "ENDF/B Summary Documentation," BNL-NCS-17541 (ENDF-201) (3rd edit. (ENDF/B-V), July 1979).
8. T. W. Burrows and N. E. Holden, "A Source List of Nuclear Data Bibliographies, Compilations, and Evaluations," BNL-NCS-50702 (Aug. 1977).
9. M. J. Martin, ed., *Nuclear Data Sheets*, Academic Press, New York, Vols. 38–40 (1983).

Oak Ridge Radiation Shielding Information Center

The Radiation Shielding Information Center (RSIC) is located at the Oak Ridge National Laboratory, Oak Ridge, TN 37830. The Center is sponsored by the Department of Energy, the Nuclear Regulatory Commission and the Defense Nuclear Agency, to specialize in acquisition of information on shielding and transport of ionizing radiation through matter. It also analyzes, selects and stores the evaluated information for subsequent dissemination to users. A large, computer based information system is operated by the Center. Computer codes for analysis of radiation transport and reactor shielding are checked out and packaged for distribution. In addition, evaluated and processed data packages, containing neutron and gamma ray cross sections and kerma factors, are prepared.(1)–(3)

The Center disseminates technology in the field of the evaluation of neutron cross sections and their uncertainties, which is important in the process of cross-section adjustment. Among the sources of uncertainty are those that enter through the measurements of neutron differential cross sections, determination of neutron spectra, instrumental effects and modeling methods. To determine the effect of each of these possible sources of uncertainties, sophisticated methods of calculation of covariance matrices have been developed, and applied to isotopes and elements of interest. The importance of this approach rests in the fact that, in reporting the measured values of cross sections, the various corrections for instrumental response tend to have systematic and highly correlated, rather than random, effects, which can be detected through the use of this methodology.[(4)–(6)] A provision was made in the Evaluated Nuclear Data File of the Brookhaven National Nuclear Data Center to accommodate these covariance matrices. They contain information on both the standard deviation and the correlation of various cross sections that can be used in evaluating reactor and shield performance.[(7)]

Recently the Center has been specializing in the production of coupled photon–neutron cross-section sets[(8)] for calculations carried out in support of the fusion program. In that program, the radiation damage caused by neutrons in the energy range above 10 MeV is of prime importance. Consequently, efforts are being made to extend the existing cross-section sets to at least 20, and preferably 40 MeV. With the introduction of intense high energy neutron sources, improved measurement techniques and codes incorporating advances in nuclear theory, this demand is expected to be successfully met.

Computer codes offered by the Center are grouped into the Computer Code Collection (CCC), and the Peripheral Shielding Routines (PSR); the latter are designed for auxiliary data processing. The CCC group includes one- and two-dimensional discrete ordinate codes, the kernel integration codes and the Monte Carlo codes. The last two groups of codes are generally three-dimensional. The codes in the PSR group serve to generate multigroup cross-section sets, exercise nuclear models, unfold experimental energy spectra, perform various optimizations, plot data and convert the results of discrete ordinate calculations to a form suitable for input to Monte Carlo calculations. The code collection of the Center contains now over 440 code packages in the CCC group, and 195 in the PSR group. Some of the most often requested codes in groups CCC and PSR are listed in Table 6. A further group of codes offered by the Center includes codes based on the Spinney removal-diffusion method, the moments method, spherical harmonics and the invariant embedding transport theory. Additional codes compute the fission product buildup, the dose due to fission product release and the stopping power for alpha and beta radiation.

Responding to the demand from different research centers for preprocessed multitable, multigroup transport theory constants, the Center has taken the lead in supplying a range of cross section sets, making use of the processing codes available at the Center. In addition, a new, much improved system for cross-section processing, called VITAMIN C, has been developed. This effort can be expected to continue in the future, whereby complex detailed evaluated data files of ENDF/B will be reduced to simpler files designed for specific applications.[(9)–(14)]

For the most used computer codes, the Center prepares user packages consisting of documented multigroup cross-section sets. Each set carries a Data Library Collection number (DLC). As the need arises, these packages are subject to revision and updating, and as the new versions appear, they are announced in the Monthly Newsletter of the Center. Table 7 is a selected list of the current offerings, which now total over 100 data packages.

In addition to codes and cross sections, the Center collects and distrib-

Table 6. Typical Code Packages Supplied by RSIC

RSIC No.	Dim.	Code Name	Description
PSR-13	0-D	SUPERTOG	Neutron X-sect. processor
PSR-52	0-D	MACK	Kerma factor generator
PSR-63	0-D	*AMPX	Neutron–gamma X-sect. processor
PSR-171	0-D	NJOY	Neutron–gamma X-sect. processor
CCC-79	0-D	ISOSHLD	Isotopic gamma source & shielding
CCC-112	0-D	SAND II	Spectra unfolding
CCC-137	0-D	RIBD II	Fission product inventory & source
CCC-233	0-D	CRYSTAL BALL	Spectra unfolding
CCC-371	0-D	*ORIGEN II	Fission product inventory
CCC-375	0-D	GALE	Gaseous & liquid effluent source
CCC-107	1-D	ETRAN	Electron/photon Monte Carlo
CCC-204	1-D	SWANLAKE	Sensitivity analysis
CCC-254	1-D	*ANISN	Discrete ordinate multigr. transport
CCC-266	1-D	ONETRAN	Discrete ordinate multigr. transport
CCC-222	2-D	TWOTRAN II	Discrete ordinate multigr. transport
CCC-377	2-D	*TRIDENT-CTR	Discrete ordinate triangular mesh
CCC-276	2-D	*DOT 3.5	Discrete ordinate multigr. transport
CCC-203	3-D	*MORSE-CG	Monte Carlo transport multi-group
CCC-187	3-D	*SAM-CE	Monte Carlo transport
CCC-289	3-D	SKYSHINE	Indirect source kernel integration
CCC-307	3-D	QUAD-CG	Kernel integration
CCC-424	3-D	SCALE-1	Out-of-core criticality & shielding

* This code is discussed more fully later in the text.

Table 7. Major Cross Section Sets Available Through RSIC

DLC No. and Name	Energy Neutron	Groups Gamma	Order of Scat-tering	Data Type	Purpose or Application
29/MACKLIB	100	0	0	Kerma factors	Dosimetry
31/FEWG1	37	21	P(3)	Cross sections	Shielding
37/EPR	100	21	P(8)	Cross sections	Fusion analysis
41/VITAMIN-C	171	36	P(3)	Cross sections	Fusion analysis
43/CSRL	218	0	P(3)	Cross sections	U sol. criticals
44/COVERX	126	0	0	Covar. matrices	CSEWG criticals
45/SENPRO	126	0	0	Sensit. profiles	CSEWG criticals
60/MACKLIB-I	171	36	0	Response funct.	Fusion analysis
75/BUGLE	47	20	P(3)	Cross sections	Shield standard

utes neutron-induced gamma ray production spectra, fluence to kerma conversion coefficients, decay schemes and radioactive decay spectra, as well as detailed output from transport calculations.

RSIC sponsors symposia on topics related to shielding, issues code documentation and publishes a newsletter. Furthermore, in cooperation with the American Nuclear Society, the Center collects and publishes, from time to time, a number of well-documented problems for use in testing computational methods of shielding analysis. These problems are then used as benchmarks.

In addition to the distribution of codes and shielding information within the United States, RSIC collaborates directly with scientists in the international shielding community, and participates in international exchange agreements.

Further information about the services offered can be obtained by contacting Robert W. Roussin, Director, RSIC, Oak Ridge National Laboratory, P. O. Box X, Oak Ridge, TN 37830, tel. (615) 574-6176.

REFERENCES

1. Betty F. Maskewitz, D. K. Trubey, R. W. Roussin, and Betty L. McGill, "The Radiation Shielding Information Center—An International Technology Resource," *Trans. ANS*, 26, 563 (1977).
2. R. W. Roussin, "Abstracts of the Data Library Packages Assembled by the Radiation Shielding Information Center," ORNL-RSIC-30 (1972).
3. B. McGill, B. F. Maskewitz, C. M. Anthony, H. E. Comolander, and H. R. Hendrickson, "Abstracts of Digital Computer Code Packages Assembled by the Radiation Shielding Information Center," ORNL/RSIC-13, Vol. IV (Jan. 1976).
4. J. D. Drischler and C. R. Weisbin, "Compilation of Multigroup Covariance Matrices for Several Important Reactor Materials," ORNL/TM-5564 (1977).

5. J. H. Marable and C. R. Weisbin, "Theory and Application of Sensitivity and Uncertainty Analysis," ORNL/RSIC-42, pp. 25–48, Oak Ridge National Lab. (1979).
6. J. D. Smith and B. L. Broadhead, "Multigroup Covariance Matrices for Fast Reactor Studies," ORNL/TM-3789 (Apr. 1981).
7. D. Rinaldis, J. M. Kellfelz, E. Kujawski, and J. H. Marable, "Evaluation of Integral Parameter Correlations and Reactor Performance Using Nuclear Data Covariances," *Trans. ANS*, 33, 859–860 (Nov. 1979).
8. J. T. Kriese, "Coupled Neutron and Gamma-Ray Cross Section Sets for Fusion Reactor Calculations," ORNL-TM-4277 (Aug. 1973).
9. W. E. Ford, C. C. Webster, and R. M. Westfall, "A 218-Group Neutron Cross-Section Library in the AMPX Interface Format for Criticality Safety Studies," ORNL/CSD/TM-4 (1976).
10. "DLC-2F: 100-Group Neutron Cross-Section Data Based on ENDF/B," RSIC-ORNL, Oak Ridge National Lab. (1975).
11. "SINEX, DLC-24: 100-Group Neutron Reaction Cross-Section Data Generated by SUPERTOG from ENDF/B," RSIC-ORNL, Oak Ridge National Lab. (1973).
12. "BUGLE: Coupled 45 Neutron, 16 Gamma Ray, P3 Cross Sections for Studies by the ANS 6.1.2 Shielding Standards Working Group on Multigroup Cross Sections," DLC-47, Oak Ridge National Lab. (1977).
13. Y. Gohar and M. A. Abdou, "MACKLIB-IV, A Library of Nuclear Response Functions Generated with the MACK-IV Computer Program from ENDF-B-IV," ANL/FPP/TM-106, Argonne National Lab. (1978).
14. J. Barhen el al., "VITAMIN-E : A Multipurpose ENDF/B-V Coupled Neutron-Gamma Cross Section Library," *Proc.* Int. Conf. Nuclear Cross Sections for Technology, NBS Special Publication 594 (1980).

Argonne National Energy Software Center

The National Energy Software Center owes its origin to a decision made in 1956 by a group of persons interested in promoting the development and use of computer programs for reactor design. Informally known as the Nuclear Codes Group, it met at the second Annual Meeting of the American Nuclear Society and decided to collect computer code abstracts submitted by members,(1) and to publish them in a suitable newsletter. The format chosen for the abstracts followed that established by document AECU-3078, issued by the Atomic Energy Commission.(2)

During the next three years, until the end of 1959, the newsletter accumulated 300 abstracts,(3) which were then edited and published by Ward Sangren at General Atomics, to be followed in 1960 by a supplement of 100 abstracts. In the same year Sangren also published *Digital Computers and Nuclear Reactor Calculations*,(4) the first book on the subject to appear on the market. It summarized the state of the art at the beginning of the 1960s, and described some early successes, such as the PDQ code, which is still in use in version eight, after many revisions and improvements.

After the Nuclear Codes Group of ANS had expanded into the Mathematics and Computation Division of the Society, a proposal was made in 1960 to establish a central computer program library in the areas of nuclear physics, reactor design and engineering. The Argonne National Laboratory undertook to serve as a central agency, and the Argonne Code Center was born. The Code Center published program summaries in condensed version in the report ANL-8040.[(5)] In due time the Code Center became the predecessor of the present National Energy Software Center (NESC). Located at the Argonne National Laboratory, 9700 South Cass Avenue, Argonne, IL 60439, it is the largest repository of computer codes for nuclear reactor design.

The NESC is responsible for editing and publishing the code abstracts, plus collecting, testing, maintaining and distributing the code packages. The center also publishes a compilation of program abstracts, ANL-7411,[(6)] and supplies the codes to qualified users. Distribution of library packages and authorizations for their use to NESC program participants reaches 1200 to 1400 copies per year out of the current NESC stock of 924 codes. Approximately 125 to 150 code packages are added to NESC code collection each year. NESC also distributes a benchmark problem book.[(7)]

Table 8 lists the principal categories of codes in the collection. Reactor development and technology accounts for about 60% of the NESC collec-

Table 8. Code Categories Carried in the NESC Collection

Cross section and resonance calculations
Neutron spectrum calculations and generation of group constants
Static design studies of reactor systems
Fuel depletion, management and economics
Space-independent kinetics of reactors
Space-dependent kinetics coupled to hydro- and thermodynamics
Radiological safety hazard and accident analysis
Heat transfer and fluid flow
Structural analysis, stress and deformation calculations
Gamma heating and shield design
Reactor system analysis
Computation of physical properties of materials
Environmental and earth sciences, geophysics, hydrology
Space sciences, analysis of orbits
Automated design of electronic equipment
Chemical analysis, mass spectroscopy, radiation chemistry
Design of particle accelerators, cyclotrons, synchrotrons
Magnetic fusion research, plasma electro-hydrodynamics
Mathematics and data handling routines
Biological and radiological studies of plants, animals and men

tion, mathematical software and computing routines constitute 15%, heat transfer and fluid flow contribute another 10%, whereas the remainder covers other topics. Most programs are written in FORTRAN, and they are often available in more than one machine-specific version. The computer models and makes represented are IBM-370, IBM-3033, CDC-6600 and 7600 and CYBER, UNIVAC-1100.[8]

The present director of the Center, Margaret Butler, has been its guiding spirit over the years. She deserves credit for the formulation of the better part of the procedures under which the Center operates. Under these procedures, NESC operates in close cooperation with its participating installations, and the responsibility for various phases of its operation is divided between the Center and its several installation representatives.

The codes handled by NESC come from various national laboratories, commercial reactor suppliers and universities. It is the responsibility of the authors to submit the codes to NESC, either directly, through the developing organization, or indirectly, through the installation representative. The installation representatives are specialists in numerical analysis who are on the staff of the given registered participating institution, be it a DOE laboratory, a university or an industrial contractor. They undertake to serve as a source of information concerning the codes available from NESC, do preliminary evaluations of codes before they are sent from the given installation to NESC, and distribute codes supplied from NESC. They are customarily made responsible for library of programs at their installations, as well as for answering the questions and inquiries received by NESC concerning the software originating at their institution.

NESC code collections form three thick volumes of code abstracts, with each abstract taking up from two to six pages. As can be expected, the popularity of the codes varies greatly from code to code, and depends on the demand for a particular type of calculation and the inherent merits of the code. Some measure of frequency of selection can be obtained from the number of references to a particular code appearing in the *Transactions of ANS*. Table 9 shows such a frequency tabulation for the most often used codes.

In addition to computer codes available for general distribution, NESC also carries codes with restricted distribution. These fall into three categories, denoted by a letter preceding the program number.

R—restricted to U.S. use. Many codes in this category are also restricted to use by U.S. citizens or nationals; some are not.

G—limited to use on U.S. government contracts only. In addition to the release form requirement, the contract identification and the signa-

Table 9. Number of Studies Reporting Use of Code in Transactions of ANS

Code Name	1982	1981	YEAR: 1980	1979	1978
*ANISN[10]	12	14	12	22	17
*COBRA-IIIC[11]	8	9	10	19	14
COMMIX-1A[12]	6	5	4	4	4
*DOT-IV[13]	6	14	8	9	13
*KENO-IV[14]	10	9	8	11	14
LEOPARD[15]	3	2	8	6	4
*MARCH[16]	6	12	1	—	—
*MCNP[17]	14	10	5	9	6
*MORSE[18]	5	14	8	7	16
NJOY[19]	1	4	3	6	—
*ORIGEN[20]	5	8	8	6	9
*PDQ-7[21]	9	6	10	13	4
*RELAP5[22,23]	10	10	9	11	12
*RETRAN[24]	6	10	11	5	4
*SIMMER-II[25]	7	8	4	1	6
SIMULATE[26]	8	8	6	7	3
SSC-L[27]	2	4	3	4	2
TRAC-PD2[28]	5	14	11	5	5
*2DB[29]	2	4	5	4	2
VENTURE-III[30]	3	4	2	1	3

* The code is discussed more fully later in the text.

ture of the contracting officer must be furnished when such programs are requested. NESC is not authorized to distribute such programs outside the United States.

S—programs distributed to a special list of designated recipients only.

Out of 924 programs in the NESC collection, 150 programs are listed in the R category, 7 in the S category, and 3 in the G category.

International exchange agreements broaden the scope of operations of the Center. Like RSIC, NESC supplies codes on request to the Computer Code Library of the Nuclear Energy Agency at Ispra, Italy,[9] and in return receives codes developed in Europe. Only a small part of the NESC holdings is restricted in distribution to domestic users, for reasons of national security. Codes originating from NESC are marked in the NEA program list with the symbol "U."

The exchange agreements for computer codes are one of the outcomes of the "Atoms for Peace" program, which also created the International

Atomic Energy Agency in Vienna, and initiated a large publication program for monographs, dealing with all aspects of the development and operation of nuclear reactors.

REFERENCES

1. Margaret Butler, *Computer Technology, Program Interchange, and Standards,* Vol. 9, *Advances in Nuclear Science and Technology,* Academic Press, New York (1976).
2. A. Radkowsky and R. Brodsky, "A Bibliography of Available Digital Computer Codes for Nuclear Reactor Problems," AECU-3078 (Oct. 1955).
3. V. Nather and W. Sangren, "Reactor Code Abstracts," *Comm.* ACM, Vol. 2 (Jan. 1959) and Vol. 3 (Jan. 1960).
4. Ward C. Sangren, *Digital Computers and Nuclear Reactor Calculations,* John Wiley & Sons, New York (1960).
5. Argonne Software Center, "USAEC Computer Program Summaries," ANL-8040 (1972), and supplements issued in following years under the title "Computer Software Summaries."
6. J. M. Brown et al., "National Energy Software Center: Compilation of Program Abstracts," ANL-7411 (rev. 1979), and Supp. 1 (1982). (In loose-leaf format, continuously updated.)
7. ANS Mathematics and Computation Div., "Argonne Code Center: Benchmark Problem Book," ANL-7416, Suppl. 2 (June 1977).
8. M. K. Butler, "National Energy Software Center—Its Program and Software Collection," *Proc.* ANS Topical Meeting, Advances in Computations, Salt Lake City, pp. 192–193, (Mar. 28–31, 1983).
9. "KWIC Index of Nuclear Programs," Nuclear Energy Agency, Ispra, Italy (Nov. 1977).
10. W. W. Engle, Jr., "A User's Manual for ANISN, A One-Dimensional Discrete Ordinates Transport Code with Anisotropic Scattering," K-1693, Computing Technology Center, Oak Ridge National Lab. (1970).
11. E. P. Coomes, E. U. Khan, T. L. George, R. L. Cheetham, and D. R. Rector, "User Manual: COBRA-WC A Version of COBRA for Single-Phase, Single and Multiassembly Thermal-Hydraulic Transient Analysis," PNL-4303, Pacific Northwest Lab. (May 1982).
12. W. T. Sha, H. M. Domanus et al., "COMMIX-1: A Three-Dimensional Transient Single-Phase Component Computer Program for Thermal Hydraulic Analysis," NUREG/CR-0785, ANL-77-96, Argonne National Lab. (Sept. 1978).
13. W. A. Rhoades, D. B. Simpson, R. L. Childs, and W. W. Engle, Jr., "The DOT-IV Two-Dimensional Discrete Ordinates Transport Code with Space-Dependent Mesh and Quadrature," ORNL/TM-6529, Oak Ridge National Lab. (1979).
14. L. M. Petrie and N. F. Cross, "KENO-IV—An Improved Monte Carlo Criticality Program," ORNL-4938, Oak Ridge National Lab. (1975).
15. R. Barry, "LEOPARD—A Spectrum Dependent Non-spatial Depletion Code for the IBM-7094," WCAP-3741, Westinghouse Atomic Power Div. (1963).
16. R. O. Wooton and H. I. Avci, "MARCH (Meltdown Accident Response Characteristics) Code Description and User's Manual," BMI-2064, Battelle Columbus Labs. (1980).
17. Los Alamos Monte Carlo Group, "MCNP—A General Monte Carlo Code for Neutron and Photon Transport," LA-7396-M, Revised, Los Alamos National Lab. (1981).

18. E. A. Straker, P. N. Stevens, D. C. Irving, and V. R. Cain, "The MORSE Code: A Multigroup Neutron and Gamma-Ray Monte Carlo Transport Code," ORNL-4585, Oak Ridge National Lab. (1970).
19. R. E. MacFarlane, R. J. Barrett, D. W. Muir, and R. M. Boicourt, "The NJOY Nuclear Data Processing System: User's Manual," LA-7584-M, Los Alamos National Lab. (Dec. 1978).
20. A. G. Croff, "ORIGEN2—A Revised and Updated Version of the Oak Ridge Isotope Generation and Depletion Code," ORNL-5621, Oak Ridge National Lab. (July 1980).
21. W. J. Cadwell, "PDQ-7 Reference Manual," WAPD-TM-678, Westinghouse Electric Corp. (1967).
22. V. H. Ransom et al., "RELAP5/MOD1 Code Manual, Volume 1: System Models and Numerical Methods, and Volume 2: Users Guide and Input Requirements," NUREG/CR-1826 (EGG-2070), EG&G Idaho (Apr. 1982).
23. "RELAP4/MOD5: A Computer Program for Transient Thermal-Hydraulic Analysis of Nuclear Reactors and Related Systems," Vols. 1–3, ANCR/NUREG/1335 (Sept. 1976).
24. J. H. McFadden, M. P. Paulsen, et al., "RETRAN-02: A Program for One-Dimensional Transient Thermal-Hydraulic Analysis of Complex Fluid Flow Systems," EPRI CCM-5, Electric Power Research Institute (1981).
25. L. L. Smith, "SIMMER-II: A Computer Program for LMFBR Disrupted Core Analysis," NUREG/CR-0453, LA-7515-M, Los Alamos National Lab. (1980).
26. D. M. Verplanck, "Manual for Reactor Analysis Program SIMULATE," RP-701-1, Yankee Atomic Electric Co. (1978).
27. A. K. Agrawal, "An Advanced Thermohydraulic Simulation Code for Transients in LMFBRs (SSC-L Code)," BNL-NUREG-50773, Brookhaven National Lab. (1978).
28. Safety Code Development Group, "TRAC-PD2: An Advanced Best-Estimate Computer Program for Pressurized Water Reactor Loss-of-Coolant Analysis," LA-8709-MS, NUREG/CR-2054, Los Alamos National Lab. (1981).
29. W. W. Little, Jr. and R. W. Hardie, "2DB User's Manual—Rev. 1," BNWL-831, Pacific Northwest Lab. (1969).
30. D. R. Vondy et al., "The Bold VENTURE Computation System for Nuclear Reactor Core Analysis, Version III," ORNL-5711, Oak Ridge National Lab. (June 1981).

Livermore Nuclear Data Center

The Lawrence Livermore National Laboratory, Box 808, Livermore, CA 94550, which is primarily a nuclear weapons design laboratory, also supports a large effort in magnetic fusion and inertial confinement fusion. To provide the necessary data for these activities, the laboratory has created and maintains collections of evaluated nuclear data for reactions induced by neutrons (ENDL, ACTL), photons (EGDL) and charged particles (ECPL). Because of laboratory needs there is necessarily an emphasis on high energy data, especially for neutron-induced reactions, but the energy range for incident neutrons considered extends from 10^{-10} to 20.0 MeV. The evaluation activity is centered in the Physical Data Group of the Theoretical Physics Division of LLNL. The work was started in 1957 by Robert J. Howerton as a one man activity.[(1)] Within a few years it was expanded to a group effort that included, in addition to collecting experimental data and producing evaluations for neutron-induced data,[2] the development of processing codes. These codes are needed to produce the calculational constants required by the neutron transport codes and the laboratory's effort in Monte Carlo code development. The latter effort is also handled by the Physical Data Group because of the intimate relationship between the Monte Carlo code (TART) and the data upon which it depends.[3] A second reason is that the Monte Carlo process is well suited for testing evaluated data against bulk experiments such as critical assemblies and 14 MeV pulsed spheres, and the quality of evaluated data cannot be assessed until such testing has been done (see Vol. 15, Part E of UCRL-50400 as an example).

The various evaluated data files are made available periodically to interested users. They may be obtained on request from the National Nuclear Data Center, Brookhaven National Laboratory, Upton, NY 11973; the Radiation Shielding Information Center, Oak Ridge National Laboratory, Oak Ridge, TN, 37830; or the Nuclear Data Section, International Atomic Energy Agency, Wagrammerstrasse 5, P. O. Box 100, A-1400 Vienna, Austria. Other inquiries should be made to Robert J. Howerton, L-71, Lawrence Livermore National Laboratory, P. O. Box 808, Livermore, CA 94550. The files that are currently available are listed in Table 10.

The system for maintaining data libraries at LLNL is designed to facilitate the retrieval of reliable, up-to-date nuclear data for design. It selects specific types of data on request and formats it in accordance with the input requirements of computer codes that will use it.

The ENDL, ECPL, ACTL and ENSL libraries are documented in the familiar, gold (more recently buff)-colored 25 volume report series, "An Integrated System for Production of Neutronics and Photonics Calculational Constants," UCRL-50400.[3]–[4] This report series constitutes the

Table 10. Nuclear Data Libraries Available at LLNL

Name	Contents of the Library
ENDL	Evaluated neutron-induced reactions
ACTL	Evaluated neutron-induced activation cross sections
ECPL	Evaluated charged-particle-induced reactions
EGDL	Evaluated photon-induced reactions
ENSL	Evaluated nuclear structure data
CDRL	Combined reaction and decay nuclear structure data
ECSIL	Experimental neutron induced nuclear data[4]

most complete and detailed documentation of a cross-section handling system in this country. Its component parts are listed in Table 11 as an example of the depth of coverage required to establish an up-to-date nuclear data base.

Table 11. Content of the UCRL-50400 Report Series: "An Integrated System for Production of Neutronics and Photonics Calculational Constants"

Volume	Report Title
I-A	ECSIL: System for Storage, Retrieval, and Display of Experimental Neutron Data, 1976
I-B	ECSX4 (Version 78-1): Conversion of Experimentally Measured Cross-Section Data from the Four Center Exchange (X-4) Format to Livermore ECSIL Format, 1978
II	A Bibliography of the Experimental Data of Neutron-Induced Interactions, 1976
III	An Index of the Experimental Data of Neutron-Induced Interactions, 1976
IV	Evaluated Nuclear Cross Section Library, 1971
V-A	CLYDE: A Code for the Production of Calculational Constants from Nuclear Data, Rev. 1, 1975
V-B	Relativistic Transformations between Center-of-Mass and Laboratory Systems for Two-Body Nuclear Reactions, 1978
VI	Tables and Graphs of Photon-Interaction Cross Sections from 1 keV to 100 MeV, 1978
VII-A	Major Neutron-Induced Interactions ($Z<55$): Graphical, Experimental Data, 1976
VII-B	Major Neutron-Induced Interactions ($Z>55$): Graphical, Experimental Data, 1976
VIII-A	Supplemental Neutron-Induced Interactions ($Z<35$): Graphical, Experimental Data, 1976
VIII-B	Supplemental Neutron-Induced Interactions ($Z>35$): Graphical, Experimental Data, 1976
IX	Thresholds of Nuclear Reactions Induced by Neutrons, Photons, Deuterons, Tritons, and Alpha Particles, 1970
X	Tabulated Experimental Data for Neutron Induced Interactions, Rev. 1, 1976
XI	Experimental Data, Indexes, and Techniques of Obtaining a Selected Set of Neutron Resonance Parameters, 1972
XII	An Atlas of Resolved Neutron Resonance Parameters, 1972
XIII	An Atlas of Unresolved Neutron Resonance Parameters, 1972
XIV	TARTNP: A Coupled Neutron-Photon Monte Carlo Transport Code, 1976
	The LLL Evaluated Nuclear Data Library (ENDL):
XV-A	Evaluation Techniques, Reaction Index, and Descriptions of Individual Evaluations, 1975
XV-B	Graphs of Cross Sections from the Library, Rev. 1, 1978

Table 11. Continued

Volume	Report Title
XV-C	Translation of ENDL Neutron Induced Interaction Data into the ENDF/B Format, 1976
XV-D	Descriptions of Individual Evaluations for Z=0-90, 1978
XV-E	Data Testing Results for the LLL Nuclear Data Library ENDL-78, 1979
XV-F	Experimental and Evaluated Elastic Nuclear Plus Interference Cross Sections for Light Charged Particles, 1980
XVI	Tabular and Graphical Presentation of 175 Neutron Group Constants Derived from the LLL Evaluated Neutron Data Library (ENDL), Rev. 2, 1978
	The Evaluated Nuclear Data File/Version B (ENDF/B):
XVII-A	Program LINEAR (Version 79-1): Linearization of Data in the ENDF/B, Rev. 2, 1979
XVII-B	Program SIGMA1 (Version 79-1): Doppler-Broaden Cross Sections in the ENDF/B, Rev. 2, 1979
XVII-C	Program RECENT: Reconstruction of Energy-Dependent Cross Sections from Resonance Parameters in the ENDF/B, 1979
XVII-D	Program GROUPIE: Calculation of Self Shielded Cross Sections and Multiband Parameters from Evaluated Data in the ENDF/B Format, 1980
XVII-E	Program EVALPLOT: Plot Data in the Evaluated-Nuclear-Data File/Version B (ENDF/B) Format, 1979
XVII-F	DOWNER (Version 79-1) Group Collapse Cross Sections and Transfer Matrices, 1979
XVIII	ACTL: Evaluated Neutron Activation Cross-Section Library, 1978
XIX	Neutron-Induced Angular and Energy Distributions: Graphical Experimental Data, 1977
XX	Bondarenko Self-Shielded Cross Sections and Multiband Parameters Derived from the LLL Evaluated-Nuclear-Data Library (ENDL), 1978
XXI-A	Maxwell-Averaged Reaction Rates (sigma.v) for Selected Reactions between Ions with Atomic Mass < 11, 1979
XXI-C	Program SIGMAL (version 79-1): Doppler-Broaden Evaluated Cross Sections in the Livermore-Evaluated Nuclear Data Library (ENDL) Format, 1979
XXII	GAMIDENT: A Program to Aid in the Identification of Unknown Materials by Gamma Ray Spectroscopy, Rev. 2, 1983
XXIII	ENSL and CDRL: Evaluated Nuclear Structure Libraries, 1981; Addendum: ENSL82 and CDRL82: the 1982 version of Evaluated Nuclear Structure Libraries ENSL and CDRL, 1983
XXIV	Thresholds and Q Values of Nuclear Reactions Induced by Neutrons, Protons, Deuterons, Tritons, ^{3}He Ions, Alpha Particles, and Photons, 1981
XXV	Omega: Cray 1 Computer Code that Controls all Operations on the LLNL Family of Evaluated Nuclear Data Libraries, 1983

REFERENCES

1. Robert J. Howerton, "Semi-Empirical Neutron Cross Sections, 0.5– 15 MeV," UCRL-5351, Lawrence Radiation Lab. (Nov. 1958).
2. Robert J. Howerton, "The Lawrence Livermore Laboratory Evaluated Nuclear Data Library (ENDL) Translated into the ENDF/B Format," UCID-16727, Lawrence Livermore Lab. (March 1975), later superseded by Vol. 15, Part C of the UCRL-50400 series.
3. E. F. Plechaty, D. E. Cullen, R. J. Howerton, and J. R. Kimlinger, "Tabular and Graphical Presentation of 175 Neutron Group Constants Derived from the LLL Evaluated Neutron Data Library (ENDL)," UCRL-50400, Vol. 16 (Jan. 29, 1975).
4. R. J. Howerton, "Tabulated Neutron Cross Sections, Part I, .001–14.5 MeV," UCRL-5226, Lawrence Radiation Lab. (May 1958), later superseded by Vols. 1, 2, 3, 7, 8, 10 and 19 of the UCRL-50400 series.

Lawrence Livermore Criticality Data Base

The most recent compilation of criticality data was prepared by Brian L. Koponen and co-workers at Lawrence Livermore Laboratory[1] as part of a much larger project aimed at creating a national computerized criticality data base. Intended for use by government and industry, the content of this data base is comprehensive, and incorporates all criticality data necessary for the design of nuclear reactors, fuel storage facilities and transportation equipment, as well as chemical reprocessing plants. Both domestic and foreign criticality data are included as far as possible.

The content of the LLL criticality data base includes over a thousand documents on experiments and their supporting calculations. It excludes, however, the results of theoretical criticality studies, and reactor design studies. The documentation consists of reports, journal articles and papers presented at society meetings and symposia authored by investigators from government, industry and universities. The abstracts of these documents are entered into the data base together with the author's name and affiliation and a description of the experimental program under which the work was performed.

As it presently stands, this data base contains information on exponential, pulsed and static, subcritical and critical assemblies for single solid or fluid configurations, arrays of fissile units and fuel element lattices. The measurements deal with systems supporting a variety of neutron energy spectra. Fast, intermediate and thermal systems are represented.

The computerization of this data base was done by means of the Lawrence Livermore MASTER CONTROL[2] computer program. It permits storage of bibliographic and numerical data, plus data handling and retrieval by on-line searches through suitable indices and key words. The intent is to make the data base useful in validation of cross-section sets and computer codes, while making use of the procedures established by the standard "Validation of Calculational Methods for Nuclear Criticality Safety," ANS 8.11.

REFERENCES

1. Brian L. Koponen, Thomas P. Wilcox, and Viktor E. Hampel, "Nuclear Criticality Experiments from 1943 to 1978. An Annotated Bibliography: Vol. 1, Main Listing; Vol. 2, Lookup Tables; Vol. 3, Subject Index," UCRL-52769, Lawrence Livermore National Lab. (Apr. 1979).
2. V. A. Hampel and J. R. Wade, "MASTER CONTROL—A Unifying Free-Form Data Storage and Data Retrieval System for Dissimilar Data Bases," *Proc.* 32nd Ann. Mtg. Am. Soc. Inf. Sci., 6, 159 (1969).

3.2 THERMODYNAMIC PROPERTY DATA COLLECTIONS

Designers participating in the development of reactors are acutely aware of the importance of obtaining good thermodynamic data in the process of assuring a successful design. For many research and development projects the difficulty of securing reliable values of materials data, and the need to evaluate their adequacy, are the prevailing factors relating to productivity.[(1)] For over a century, the experimental data in this area have been used to supply a skeleton around which tabulations and various thermodynamic charts have been constructed. In the process, gaps in the information have been filled by spline fits and interpolations, and much of the detail of the original data has been concealed.

A problem specific to the presentation of experimental results of thermodynamic data concerns the manner in which the data subsequently will be used. If the data are intended for machine calculations, the user often starts from tables, and makes his own fits to the tabular form. He then develops his own thermodynamic-property subroutine, to generate point property values needed in reactor design calculations. Alternately, he may computerize the various equations of state generated in the process of producing the tables in the first place. In either case, the fitting process is carried out twice: from measured point values of thermodynamic property to tables, and from tabular values to computer-generated point values demanded by engineering design.

The conventional manner of publishing thermodynamic data only through tabulations and charts is no longer adequate for the needs of the reactor designer, although it still supplies him with a means of checking his thermodynamic-property computer subroutines. One can expect that in response to demands of present and future users, the form of presentation of thermodynamic data will change, and the experimental data will become more accessible. Thermodynamic packages containing both the measured data with its experimental uncertainties and the computer subroutines mechanizing the corresponding equations of state will become the rule.

Thermodynamic input data used in the development and design of nuclear reactors originates from a great variety of sources. Some of it reaches the designer through the traditional channels of engineering information: handbooks, data compilations derived from technical journals and the resources of engineering companies active in the field. The rest is derived from directed research, and appears in topical technical reports. The following sections discuss specifics of the existing collections of thermodynamic data relevant to the field of nuclear engineering.

Generally speaking, thermodynamic data collections remain quite fragmented, and appear in tabulations scattered through many publications. Some years ago, an effort was initiated under the auspices of the National Academy of Sciences and the National Bureau of Standards to bring some coherence to this field, during a conference held in Warrenton, VA. On that occasion, the need for a computerized national thermodynamic property data base was discussed.[(2)]

The thermodynamic data for coolants are the most complete, compared to other materials, and are based on existing formulations of equations of state, and are tabulated.[(3)–(5)] Nuclear energy conversion systems are commonly based on the closed loop Rankine or Brayton cycle, although the NERVA reactor, developed in the 1960s for space shuttle missions, used the open Brayton cycle. The working fluids include light water, heavy water, CO_2, He, H_2, Na or NaK, molten salts and organic compounds (waxes).

Energy balances derived from the measurements of thermodynamic state point values in an operating power reactor have been used to verify the values of energy release in fission. In one study[(6)] an operating power reactor was used as a flow calorimeter to determine the total energy release during several reactor operating cycles. Corresponding isotopic changes in the heavy metal inventory in the core were then obtained by mass spectrometric analysis of the composition of the discharged fuel, and the number of fissions was determined. The ratio of these two quantities gave an independent verification of the basic nuclear engineering constants.

REFERENCES

1. Lewis M. Branscomb, "Improving R&D Productivity: The Federal Role," keynote remarks at the workshop "Towards a National Science and Technology Data Policy," held on Apr. 14, 1983, Washington, DC, *Science*, 222, 133–135 (Oct. 14, 1983).
2. J. Jedruch, "Handling of Thermodynamic Data by Industrial Users and the Need for a

National Thermodynamic Data Bank," prepared remarks, pp. 16–18, "Report of the Conference on Thermodynamics and National Energy Problems," National Academy of Sciences, held at Airlie House, Warrenton, VA (June 10–12, 1974).
3. Joseph H. Keenan, Frederick G. Keyes, Philip G. Hill, and Joan G. Moore, *Steam Tables*, John Wiley & Sons, New York (1978). Both English and SI unit editions are available.
4. N. B. Vargaftik, *Tables on the Thermophysical Properties of Liquids and Gases in Normal and Dissociated States*, translated from Russian, John Wiley & Sons, New York (1975).
5. W. C. Reynolds, "Thermodynamic Properties in SI, Graphs, Tables and Computational Equations for Forty Substances," Department of Mechanical Engineering, Stanford University, Stanford, CA, 1979.
6. J. Jedruch and R. J. Nodvik, "Verification of Energy Release in Fission from Power Plant Calorimetrics and Isotopic Composition of Spent Fuel," *J. Nucl. Energy*, 27, 195–225 (1973).

Computerization of Thermodynamic Calculations

Thermodynamic data are involved in the design of energy conversion devices based on nuclear fission or fusion at various stages. The data are used to characterize the properties of materials appearing in the solid, liquid or gaseous phase. Their principal use, of course, is in the analysis of energy conversion in Brayton or Rankine cycles. They are also an important factor in the analysis of accident conditions involving possible changes of phase, or energy releases in chemical reactions. They play a part in the manufacture of new, and the reprocessing of spent fuel. In normal operating conditions, interest is focused primarily on the thermodynamic properties of water, liquid metals and salts, as well as gases.

A complete nuclear design places unique demands on thermodynamic data by requiring extrapolations of temperature and pressure for postulated accident conditions. Nuclear engineering brings to the discipline of thermodynamics the routine use of advanced numerical techniques for evaluation of reaction kinetics, and efficient computerized handling of large blocks of data.

Although a significant part of the nuclear engineering data base, particularly in the area of cross sections, nuclear data and abstract indexes, is already stored on a magnetic medium, large portions of data vital to this field are still in the early stages of transfer to on-line, computer-readable files. This is primarily true of thermodynamic property data, knowledge of which is essential for the successful development of nuclear energy conversion cycles. The range of temperatures covered by a given energy conversion cycle is inversely proportional to the heat capacity of the working fluid, and to the inventory of coolant allowed in the reactor core. The coolant inventory is restrained in turn by considerations of nuclear criticality and demands for pumping power required to overcome the hydraulic resistance of the core. These often conflicting requirements of thermodynamics, hydrodynamics and neutronics of the core must be reconciled through the calculations required for reactor design.

The internal organization of large design codes has been characterized over the years by a trend toward increased coupling of the thermo-hydraulic and the neutronic aspects of calculation. Inclusion of the best thermodynamic formulations in these codes is mandatory. In computerized thermal design calculations, the values of the thermodynamic state points are computed ad hoc from equations of state incorporated into the program in the form of either functions or subroutines. In addition to accuracy, high computing efficiency is required by these computing formulations, in view of their heavy use during the execution of the program. Examples of such subroutines will be given in the sections dealing with specific coolants.

Several general computer programs[(1)–(4)] have already been developed for computing thermodynamic properties in the course of nuclear projects. Heat capacity, enthalpy, entropy and free energy are computed, fits of functions to empirical equations are performed, and heats of formation and equilibrium constants are calculated as a function of temperature. In addition, material-specific formulations for materials other than water are becoming available. In the following sections the existing thermodynamic data collections pertinent to nuclear engineering will be reviewed, and computer-oriented formulations of thermodynamic equations of state will be identified.

Analyses of accident conditions in a reactor undergoing a nuclear transient put severe demands on the handling of thermodynamic data for all constituents of the reactor core. High rates of change of temperature and pressure, changes in phase and the expulsion of core components have to be predicted for conditions of temperature and pressure well beyond the limits of the presently available measurement techniques. In the extrapolations required, generalized compressibility, enthalpy and entropy relations are heavily relied on to fill gaps in the existing data.[(5)] For analyses of this type, equations of state are required giving pressure as a function of energy and volume for input to hydrodynamic calculations, to predict the time variation of geometry and of thermal expansion.[(6)] Equations of state giving temperature as a function of energy and volume are required for calculation of the Doppler coefficient of fuel. It is particularly difficult to analyze situations where melting components of the core interact with liquid coolants on a time scale in which heat transfer coefficients and viscosities of the partially molten components become very important.

The large amount of effort spent in formulating and computerizing the thermodynamic and hydraulic relations involved in transient accident analysis is evident from the number of computer codes that have been developed during the last decade. They are discussed in detail in a later section.

REFERENCES

1. M. H. Fontana, "ONEPHASE: A Simplified FORTRAN 63 Program for Computing Tables of Thermodynamic Data for Compounds in One Phase," ORNL-4307, Oak Ridge National Lab. (1968).
2. M. H. Fontana, "MONAT: A FORTRAN 63 Program for Computing Thermodynamic Properties of Monatomic Ideal Gases," ORNL-4305, Oak Ridge National Lab. (1968).
3. M. H. Fontana, "TCDATA: A FORTRAN Program for Computing Thermodynamic Data of Compounds in Any Phase State," ORNL-4303, Oak Ridge National Lab. (1968).
4. Bonnie J. McBride and Sanford Gordon, "FORTRAN IV Program for Calculation of Thermodynamic Data," NASA-TN-D-4097, Lewis Research Center, Cleveland, OH (Aug. 1967).
5. John Graham, *Fast Reactor Safety*, Academic Press, New York (1971).
6. D. Miller, "A Critical Review of the Properties of Materials at High Temperature and Pressure Significant for Fast Reactor Safety," ANL-7120, Argonne National Lab. (1965).

Heavy Water and Tritium Compound Properties

The thermodynamic properties of heavy water could first be measured in 1933, when N. G. Lewis and R. T. MacDonald separated the first sample of pure heavy water. Early studies of nuclear chain-reacting assemblies established an interest in the properties of heavy water for nuclear technology. Indeed, German research in this area during World War II concentrated on heavy water in the belief that it was the only neutron-moderating material that would enable a chain reaction to take place. Interest in heavy water as a reactor coolant and moderator has continued in the United States in connection with defense and power reactor demonstration projects, and in Canada in support of the development of the successful CANDU line of reactors.[(1),(2)] Evaluations and tables[(3)] were prepared and used in the design of the Carolinas Virginia Tube Reactor (now defunct). The temperature range covered by these tables was from 39°F to 700°F, and pressures were from 0.068 psia to 3131.7 psia. Enthalpy, entropy, density, viscosity, thermal conductivity and specific heat were tabulated.

The data presently available on heavy water were evaluated and systematized in a recent Russian work.[(4)] Compressibility of liquid and gaseous heavy water was examined in connection with this evaluation, as well as phase transition curves. Critical parameters and specific heats in the solid, liquid and gaseous phase were tabulated. The existing data on the latent heats of fusion and vaporization were critically examined, and the thermal conductivity and viscosity of liquid and gaseous heavy water were reviewed.

Additional interest in the thermodynamic properties of higher isotopes of hydrogen is centered in the fusion programs[5] that look to heavy water as a fuel. Deuterium and tritium are the two fuel components for the most promising fusion reaction.

REFERENCES

1. J. Kirshenbaum, *Physical Properties and Analysis of Heavy Water*, McGraw-Hill, New York (1951).
2. E. Whalley, "The Thermodynamic and Transport Properties of Heavy Water," *Proc.* Conf. Thermodyn. Tr. Prop. Fluids, London (1958)
3. A. Bishop and P. A. Nelson, "Transport and Thermodynamic Properties of Saturated and Compressed Heavy Water," CVNA-58, Westinghouse Electric Co., Pittsburgh (1961).
4. Ya. Z. Kazavchinskii et al., *Heavy Water, Thermophysical Properties,* translated from Russian by the Israel Program for Scientific Translations, Jerusalem (1971).
5. Harold H. Garber, "Tritium Properties and Related Quantities Used in the TFTR Conceptual Design," WFPS-TME-004, Westinghouse Electric Co., Fusion Power Systems Dept. (Sept. 1975).

Light Water Properties

Since 1929 the thermodynamic properties of water have been the subject of concerted study supported by the American Society of Mechanical Engineers. The society participates in the International Association for the Properties of Steam, and publishes the skeleton tables.[1]

In an extension to the work done under the auspices of ASME and with ASME's permission, the properties of compressed water were reexamined and graphed on expanded scales, in the ranges of temperature and pressure of interest to pressurized water reactor design, by T. C. Tsu and D. T. Beecher of Westinghouse Research Laboratories.[2]

Following the early use of tabulations and charts of thermodynamic properties of water, the industry made a transition to computerized formulations based on the 1967 ASME Steam Tables,[1] with additions, modifications and improvements incorporated later as the need arose. The formulations of thermodynamic functions of water intended for computer use are based, as a rule, on the partial derivatives at constant pressure or at constant volume of either the Gibbs function (the specific free enthalpy) or the Helmholtz function (the specific free energy). A detailed description of the computational procedures and a complete FORTRAN listing were given in an ASME 1968 booklet by R. B. McClintock and G. J. Silvestri.[3] Later improvements in the thermodynamic formulations for water contained in these computer subroutines were described in ASME journals,[(4)–(7)] and a "Steam Properties Package," consisting of the subroutines plus the descriptive matter, is available from Prof. Richard W. Counts of the Department of Chemistry, Indiana University, Bloomington, IN 47405.

The thermodynamic properties incorporated in the computer functions are computed by a name call from the main program. The properties available as a function of two input arguments are: specific enthalpy, specific entropy, Prandtl Number, specific heat, specific volume, steam quality, thermal conductivity and viscosity. In addition, each one of them, plus the temperature and the pressure, is available as a function of a single argument on the saturation line. Generally, the range of pressures covered extends from 0 to 1000 bars, for temperatures in the range from 0.01° C to 800° C. Table 12 lists the functions and their arguments.

As part of the quality assurance program, the equations of state for

Table 12. Computer Functions for Thermodynamic Properties of Water

Property	Property Region	Function	Input Arguments
Specific enthalpy	Saturated liquid	H=HSL(T)	T
	Saturated vapor	H=HSV(P,T,S,V)	P
	Liquid	H=HCL(P,T,S)	P,T
	Liquid	H=HSSICL(P,S,T)	P,S
	Steam/Gas	H=HSS(P,T,S,V)	P,T
	Steam/Gas	H=HSSISS(P,S,T,V,X)	P,S
Specific entropy*	Saturated Liquid	S=SSL(T)	T
	Liquid	S=SSSICL(P,H,T)	P,H
	Steam/Gas	S=SSSISS(P,H,T,V,X)	P,H
Prandtl Number	Liquid	PR=PRLIQ(P,T)	P,T
	Steam/Gas	PR=PRSTM(P,T)	P,T
Pressure	Saturation line	P=PSL(T)	T
	Saturated vapor	P=PSV(S)	S
Specific heat	Liquid	CP=CPL(P,T)	P,T
	Steam/Gas	CP=CPV(P,T,V)	P,T
Specific volume**	Saturated liquid	V=VSL(T)	T
	Liquid	V=VCL(P,T)	P,T
Temperature***	Saturated liquid	T=TSHL(H)	H
	Saturation line	T=TSL(P)	P
Thermal conductivity	Liquid	C=CONDL(P,T)	P,T
	Steam/Gas	C=CONDV(P,T)	P,T
Viscosity	Liquid	VISC=VISL(P,T)	P,T
	Steam/Gas	VISC=VISV(P,T)	P,T

Notes
* Can be also obtained from HSV, HCL, HSS.
** Can be also obtained from HSV, HSS, HSSISS, CPV, SSSISS.
*** Can be also obtained from HSV, HSSICL, HSSISS, SSSICL, SSSISS.

water and steam, derived from the ASME formulations, have been subjected to extensive testing,[8] to assure their validity and to determine how closely they agree with the values of the thermodynamic properties contained in the skeleton tables of ASME. The deviations are required to fall within close tolerances. Checking and generation of three-dimensional plots of the property in question, as a function of temperature and pressure, are done by service computer programs developed specifically for this purpose.

In the years 1976 and 1977 ASME published the Mollier Chart and Steam Tables in SI units.[9],[10]

The formulation described above is used extensively in the industry, and is considered standard, and is efficient when incorporated in larger computer programs. Nevertheless, it has some shortcomings due to the discontinuous nature of its equations. To remove these discontinuities, an effort was initiated at the National Bureau of Standards by Lester Haar and John S. Gallagher, in cooperation with George S. Kell of the National Research Council of Canada, to produce a new standard for the properties of light water. A new relation between the Helmholtz free energy, the density and the temperature of water was formulated as an equation of state in a computer-programmable form.[11]

Although the new formulation of the equation of state of water is superior in accuracy to the preceding version, it takes ten times as much computer time to generate values of state points. Consequently, there are at present no plans for it to replace the 1967 formulation for industrial use on which the current ASME Steam Tables and computer subroutines are based.

In 1982 I. Z. Kuck of the University of Arizona prepared for NRC a set of formulations of thermodynamic properties[12] of water for use in computer simulation of power plants. Another set of such subroutines, suitable for use on a minicomputer, is available from the Center of Energy Studies of the University of Texas at Austin.

In 1983 ASME published a new edition of the Steam Tables,[13] based on the interim 1982 formulation of steam properties. The edition includes properties of heavy water for the first time.

In 1982 another library of FORTRAN subroutines, named STP[14] was developed by researchers at the Idaho National Engineering Laboratory for the computation of thermodynamic properties of steam. Starting from the true saturation envelope of Keenan and co-workers, the Idaho analysts simultaneously evaluated a large set of water properties, including isothermal compressibility and the speed of sound. Evaluation of the saturation envelope by the continuation method assured subroutines of an exceptionally high degree of mathematical and thermodynamic consistency, avoiding the usual difficulties at the critical point.

REFERENCES

1. "Thermodynamic and Transport Properties of Steam," American Society of Mechanical Engineers, New York (1967).
2. T. C. Tsu and D. T. Beecher, "Thermodynamic Properties of Compressed Water," Westinghouse Research Lab. Pittsburgh, PA (1957).
3. R. B. McClintock and G. J. Silvestri, "Formulations and Iterative Procedures for the Calculation of the Properties of Steam," American Society of Mechanical Engineers, New York (1968).
4. R. B. McClintock and G. J. Silvestri, "Some Improved Steam Property Calculation Procedures," *Journal of Engineering for Power, Trans. ASME* (Apr. 1970).
5. R. C. Hendricks, R. B. McClintock, and G. J. Silvestri, "Revised Representations for the Viscosity of Water and Steam and New Representations for the Surface Tension of Water," *Journal of Engineering for Power, Trans. ASME*, Series A, Vol. 99 (1977).
6. J. Kestin, "Thermal Conductivity of Water and Steam," *Mechanical Engineering*, 100, 46–48 (Aug. 1978).
7. J. V. Sengers, "Dielectric Constant of Water and Steam," *Mechanical Engineering*, 101, 44–45 (Sept. 1979).
8. K. V. Moore, "ASTEM: A Collection of FORTRAN Subroutines to Evaluate the 1976 ASME Equations of State for Water and Steam and Derivatives of These Equations," ANCR-1026, Aerojet Nuclear Company (1971).
9. J. H. Potter, ed., "Steam Charts, SI (Metric) and U.S. Customary Units," American Society of Mechanical Engineers, New York (1976).
10. "ASME Steam Tables in SI (Metric) Units for Instructional Use," American Society of Mechanical Engineers, New York (1977).
11. J. Kestin, J. V. Sengers, and R. C. Spencer, "Progress in the Standardization of the Properties of Light and Heavy Water," *Mechanical Engineering*, 105, 72–73 (Mar. 1983).
12. I. Z. Kuck, "Thermodynamic Properties of Water for Computer Simulation of Power Plants," Department of Nuclear Engineering, University of Arizona (May 1982).
13. "ASME Steam Tables—Fifth Edition," American Society of Mechanical Engineers, New York (1983).
14. F. Aguilar, P. G. Tuttle, and A. C. Hunter, "STP: A Mathematically and Physically Consistent Library of Steam Properties," *Nucl. Sci. Eng.*, 81, No. 3, pp. 467–471 (July 1982).

Properties of Gases Used as Reactor Coolants

The thermochemical properties of a large number of gases have been collected in the JANAF Thermochemical Tables,[1] issued in connection with a rocket project sponsored jointly by the Army, the Navy and the Air Force. The tables, over one thousand pages in volume, were issued as part of the National Standard Reference Data System of the National Bureau of Standards. Although initially developed in support of the development of the chemical rocket propulsion industry, the tables have found wide utilization in those projects of the nuclear industry that need the thermodynamic properties of gases at high temperature and pressure, such as high temperature gas-cooled reactors or nuclear space propulsion projects.

The development of the JANAF Tables was a group effort involving some 20 investigators from the Dow Chemical Thermal Research Laboratory, and extending over a period of ten years, from 1960 to 1970. The working group included thermodynamicists, technological users of the data and government sponsors of research and development. The meetings of the generators and the users of the data for prepublication reviews, created a technical group which survived beyond the termination of the project, and continues to exert a beneficial influence on other projects in this area.

The JANAF tables were initially distributed in the form of loose-leaf sheets, which were eventually collected, photographically reduced to a smaller format, and issued under one cover.[(1)–(2)] Computerization of access to this data base was planned early, but the transfer of these important tables to a magnetic medium has been delayed due to the scaling down of the JANAF project activity.

Experience with gases as reactor coolants has been obtained in the United States mainly for hydrogen,[(3)–(5)] nitrogen and helium.[(6)] Carbon dioxide, although popular as reactor coolant in England and France, has been used only to a limited extent in the United States.[(7)] The use of gases as coolants commonly involves high temperature devices, and in the case of the NERVA rocket engine, the temperature range also extended into the cryogenic region (2500°K to 15°K). When hydrogen is used in the cryogenic range, transformations from parahydrogen to orthohydrogen (nonaligned and aligned spins) must also be considered in treating the thermodynamic properties. The high operating temperatures of the gas-cooled reactors lead to higher energy conversion efficiencies than are possible with reactors cooled with water. These advantages, however, have to be balanced against the exacerbation of problems with the strength and endurance of materials used in the construction of heat exchangers and turbines.

During reactor design calculations the thermodynamic properties of gases are commonly recomputed ad hoc from the equations of state by suitable computer subroutines,[(4)–(6)] rather than obtained from a table lookup operation. The state properties of interest are the enthalpy, entropy, specific volume, ratio of specific heats, sonic velocity, thermal conductivity and absolute viscosity. An example of argument lists for the TAB[(5)] gas property subroutines is given in Table 13. Such subroutines are capable of generating between 25,000 and 150,000 values of state property per minute on the CDC-6600 computer.

Properties of gaseous carbon dioxide are available in the form of tables

Table 13. Argument List for the TAB Gas Property Programs

TABHP(H, P, Q, T, V, GA, CP, SV, C, VIG, VIL, VG, VL)

Input	Output	Definition	Units
H		Enthalpy	Btu/lb
P		Pressure	lb/in^2
	Q	Quality	—
	T	Temperature	°R
	V	Specific volume	in^3/lb
	GA	Specific heat ratio	—
	CP	Specific heat	Btu/lb-°R
	SV	Sonic velocity	in/sec
	C	Thermal conductivity	Btu/in-sec-°R
	VIG	Absolute viscosity	lb/in-sec
	*VIL	Liquid absolute viscosity	lb/in-sec
	*VG	Gaseous specific volume	in^3/lb
	*VL	Liquid specific volume	in^3/lb

* Apply only in the two-phase region.

TABTP(T, P, H, V, GA, CP, SV, C, VIG, ICE)

Input	Output	Definition	Units
T		Temperature	°R
P		Pressure	lb/in^2
	H	Enthalpy	Btu/lb
	V	Specific volume	in^3/lb
	GA	Specific heat ratio	—
	CP	Specific heat	Btu/lb-°R
	SV	Sonic velocity	in/sec
	C	Thermal conductivity	Btu/in-sec-°R
	VIG	Absolute viscosity	lb/in-sec
	ICE	Melting line index	—

and charts prepared in connection with the Marine Gas Cooled Reactor Program.(7) A computer program for the evaluation of thermodynamic and transport properties of interest was developed for digital calculations as well.

REFERENCES

1. D. H. Stull and H. Prophet, eds., ''JANAF Thermochemical Tables,'' 2nd edit., National Standard Reference Data System, National Bureau of Standards (June 1971).

2. M. W. Chase, J. L. Curnutt, H. Prophet, R. A. McDonald, and A. N. Syverud, "JANAF Thermochemical Tables, 1975 Supplement," *Journal of Physical and Chemical Reference Data*, 4, No. 1 (1975).
3. R. W. Patch, "Thermodynamic Properties and Theoretical Rocket Performance of Hydrogen to 100,000° K and 1.01325×10^8 N/m^2," NASA Lewis Research Center (1971).
4. B. L. Pierce, "Para-Hydrogen Codes," WANL-TME-1478, Westinghouse Astronuclear Lab. (1966).
5. O. A. Farmer, T. E. Springer, and B. B. Fisher, "FORTRAN IV Hydrogen Property Tabular Codes," LA-3381, Los Alamos Scientific Lab. (Oct. 1965).
6. B. L. Pierce, "The Thermodynamic and Transport Properties of Helium and Nitrogen," WANL-TME-1753, Westinghouse Astronuclear Lab. (1968).
7. L. H. Chen, "Thermodynamic and Transport Properties of Gaseous Carbon Dioxide," GA-1038, General Dynamics Co. (Aug. 1959).

Liquid Metals Used as Reactor Coolants

The thermodynamic properties of liquid sodium were used extensively in the design work for the Enrico Fermi fast reactor. The Fast Flux Test Facility (FFTF) constructed at Hanford, WA, and the Clinch River Liquid Metal Fast Breeder plant use liquid metal coolants in their design.

A rather complete review and reevaluation of properties of sodium was given in 1960 by Makansi, Selke and Bonilla(1) for phase regions away from the critical point, and estimates of critical point values were reported by Grosse(2) in 1961, and Dillon(3) in 1965. Thus, enough data were available for design of liquid-metal-cooled reactors at normal operating conditions.

Another useful compilation of thermophysical properties of cesium, rubidium, potassium, sodium and lithium was prepared in 1968 by four investigators at the Aerojet General establishment at San Ramon, CA.(4) Enthalpy, entropy, thermal conductivity, electrical conductivity and viscosity were presented in cgs units in tables and graphs for the liquid, vapor and gaseous forms of these metals. Suitable extrapolations were made to cover gaps in the data. Equations of state, and a discussion of the rationale used in deriving them, were also included in the text. Cesium and rubidium are not considered as coolants for power reactors on account of their high cost. They have been considered, however, as coolants for various space applications, where the small overall mass of the system reduces the importance of the cost factor.

The thermodynamic data for sodium, evaluated at the Argonne National Laboratory,(5) are the set now most frequently used by LMFBR design projects. The data have been entered on a magnetic medium, and are also available in the form of a computer subroutine. One example of the argument list of such a subroutine is given in Table 14. The reference quoted also gives the FORTRAN listing of the subroutine.

Table 14. Argument Lists of Subroutines for Sodium Properties

Phase Region	Subroutines
Subcooled	SUB(H, P, T, R, DVDHP, DVDPH)
Saturated	STAM(P, HF, HG, S, VF, VG, DHFDP, DHGDP)
Superheated	SUP(H, T, P, R, DVDHP, DVDPH)

Where:

H = Enthalpy, HF liquid, HG gas
S = Entropy
P = Pressure
T = Temperature
R = Density
C = Specific Heat
V = Specific Volume, VF liquid, VG gas
DVDPH = $(dV/dP)_H$ Partial derivatives with respect to pressure
DVDHP = $(dV/dP)_P$ Partial derivatives with respect to pressure
DHFDP = (dH_F/dP) Partial derivatives with respect to pressure
DHGDP = (dH_G/dP) Partial derivatives with respect to pressure

Another set of high temperature thermodynamic properties of sodium was prepared more recently at Hanford Engineering Development Laboratory as a result of an evaluation carried out by A. Padilla.[(6)] The availability of this set hopefully will put to rest all concern over the adequacy of sodium data in thermodynamic conditions of hypothetical LMFBR accidents.

REFERENCES

1. Munzer Makansi, W. A. Selke, and C. F. Bonilla, "Thermodynamic Properties of Sodium," *J. Chem. Eng. Data*, 5, No. 4, 441–452 (Oct. 1960).
2. A. V. Grosse, "The Temperature Range of Liquid Metals and an Estimate of Their Critical Constants," *J. Inorg. Nucl. Chem.*, 22, 23 (1961).
3. I. G. Dillon, "Critical Temperature and Densities of Alkali Metals," ANL-7025, Argonne National Lab. (Aug. 1965).
4. P. Y. Achener, W. V. Mackiewicz, D. L. Fisher, and D. C. Camp, "Thermophysical and Heat Transfer Properties of Alkali Metals," AGN-8195, Vol. 1, Aerojet-General Corporation, San Ramon, CA (1968)
5. G. H. Golden, J. V. Tokar, and D. Miller, "Thermophysical Properties of Sodium—Recommended Values," *Reactor and Fuel Processing Technology,* 11, No. 1, 27–48 (Winter 1967–1968).
6. A. Padilla, "High-Temperature Thermodynamic Properties of Sodium," HEDL-TME-77-27, Hanford Engineering Development Lab. (Feb. 1978).

4. PROPERTY AND PERFORMACE DATA BASES

4.1 MATERIAL PROPERTY DATA CENTERS

Concurrent with the growth of the nuclear industry, we have witnessed the rise of the digital computer, and its associated magnetic storage devices. The progress of these two fields has been closely linked from their very inception, and it could be argued that the development of either one would have been much impeded without the rise of the other. Most of the extensive nuclear calculations would have been impossible without the digital computer, while, on the other hand, the demand for computing capability and the financial resources committed to secure this capability assured the necessary base for the development of computers. It was a classic example of symbiosis.

The use of computers, in turn, required a continuous transfer of numerical input data to the magnetic medium, initially on a modest scale, but increasing with time, and now extending much beyond the simple entry of input data for numerical calculations. It presently embraces the handling of a voluminous output of numerical calculations and extensive storage of experimental information. The transfer of data to the magnetic medium greatly facilitates its access and retrieval, and relieves some of the hardening of the arteries that was becoming evident in the old information system, which depended on the transfer of data by hard copy.

Once the content of a data collection is transferred to computer files and mounted for access in the form of rotating disk storage, the data can be accessed randomly not only at the host installation, but also from remote locations equipped with access to the tele-communications network. Unit conversion can be included in the data-handling procedures, graphical manipulation becomes possible, and integration of downloaded data with the design algorithms of local computer software is a practical possibility. File inversion and profile matching are also among the possibilities of data handling.

An effort is currently afoot to establish an interconnected network of remotely accessible materials data bases, all sharing common accessing and formatting protocols. The initiative was provided by cooperative efforts of the National Bureau of Standards, Oak Ridge National Laboratory and CODATA, the Committee on Data for Science and Technology of the International Council of Scientific Unions. Workshops and meetings were held, the last in November of 1982, to formulate a policy that would consolidate the existing collection of some 60 unconnected, publicly available specialized data bases, and provide them with an efficient system of access and searching. This approach was chosen by consensus, in preference to the establishment of a single data base in one location.

The system currently under consideration would build on the experience gained from operation of the Technology Information System, now in place at the Lawrence Livermore National Laboratory. It uses a PDP 11/70 computer, operating under a UNIX-7 system, to serve as an intelligent gateway, shunting incoming requests for information to the appropriate data base accessed through an INGRES-6.3 relational data base manager. No translation, or reformatting of data, is required of users of this system.

Only a small fraction of materials property data tabulated in handbooks and publications of testing laboratories is presently available for on-line access. To increase data availability, the Materials Properties Council in May 1983 formed a corporation for the National Cooperative Materials Property Data Network. Data base standards for materials properties will be developed under the auspices of ASME, the National Society of Corrosion Engineers and the American Society of Metals. A good part of this work is directly applicable to nuclear engineering area.

In the sections that follow, partially or fully computerized data bases containing materials properties relevant to nuclear engineering will be discussed in more detail.

Purdue University Materials Properties Data Bases

Data centers dedicated to the collection, evaluation and dissemination of information on structural materials and coolants of interest to nuclear engineering are in various stages of organization. At present the most advanced one is the Center for Information and Numerical Data Analysis and Synthesis (CINDAS) at Purdue University, West Lafayette, IN. Its collections are of great importance to the design of the heat transfer part of nuclear power systems. CINDAS operates the Thermophysical and Electronic Properties Information Analysis Center (TEPIAC), which is one of the Department of Defense Information Analysis Centers. It is located on the campus of Purdue University, 2595 Yeager Road, West Lafayette, IN 47906, tel. (317) 494-6300. Its director is Dr. C. Y. Ho.

The Center has been in operation since 1957, and was originally known as the Thermophysical Properties Research Center (TPRC). It owes its origin and growth to the efforts of the late Prof. Y. S. Touloukian. The Center engages in scanning over 3500 technical and scientific journals that report the results of measurements of materials properties. Including technical research reports and theses, there are more than 9300 input sources for the computer-searchable files of the Center.

The distribution of information contained in the data bank of TPRC to users was carried out between 1957 and 1966 in the form of periodic mailing of loose-leaf sheets to subscribers of the TPRC Data Book.[1] Starting in 1967, the collections of TPRC were published in bound volumes.[2]–[4]

Included in the collections of the Center are data and information on thermophysical, thermoradiative, electronic, electrical, dielectric, optical and magnetic properties of all technologically important materials. Properties considered include the thermal conductivity, diffusivity, viscosity, expansion, radiative constants, specific heat, electrical resistivity, etc., of metals and alloys. All these properties are of interest to nuclear engineers, particularly in the case of metals, ceramics and graphites used in reactor construction. The property content of the Purdue data base is outlined in Table 1.

After the Center was renamed CINDAS in 1973, the computerization of its collections was begun. The aim is to develop and establish a computerized on-line interactive Materials Properties Numeric Data System for rapid computer retrieval and dissemination.

Following the extraction of numeric data from the scientific and technical literature, gross errors are eliminated, and the data are converted to as uniform a format as possible. All data included in the Center's data bank are referenced to the original work, and no secondary sources are used. Although as a rule the data are entered into the CINDAS data base as extracted, the critical evaluation of entries has been carried out for 17 properties of selected groups of materials. Where gaps exist in the data and systematization of existing collections allows interpolation, empirical procedures or detailed theoretical calculations are used to fill the voids. Depending on the available information, simple tabulations of isolated values, detailed tabulations with generating equations, or plots showing the concordance of different values are provided, see Fig. 5.

The Material Properties Bibliographic Data System contains more than 250,000 bibliographic citations, for more than 67,000 materials, identified in the materials directories with publication dates covering the period from 1850 till the present. The storage system for this data bank is fully

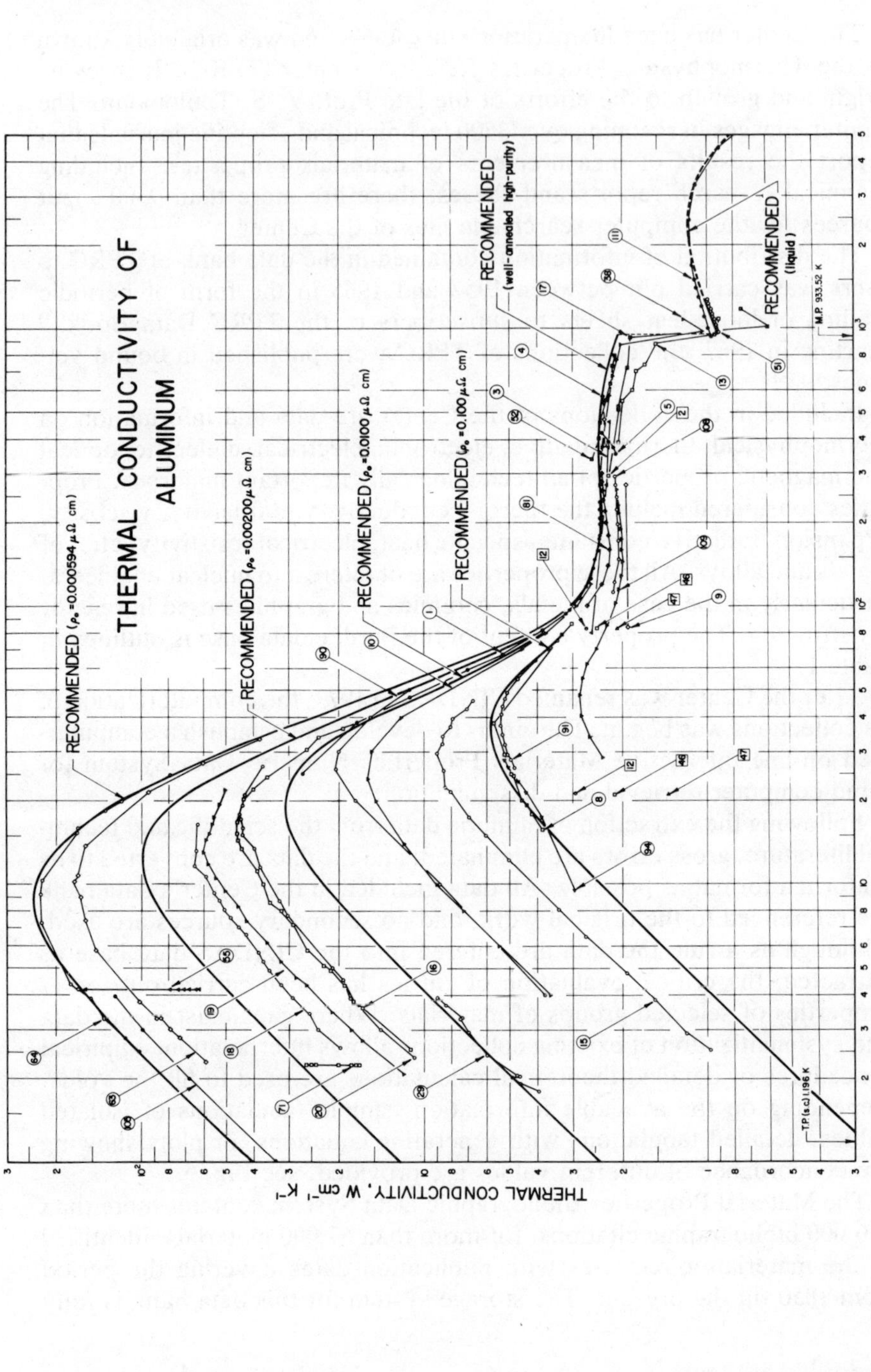

Figure 5. An example of aluminum property data from CINDAS. (Reprinted by courtesy of Thermophysical Properties Research Center)

Table 1. Proprerty Coverage of CINDAS Data Base

Thermophysical Properties	
Absorptance	Thermal Conductivity
Absorptance/Emittance Ratio	Thermal Contact Resistance
Accommodation Coefficient	Thermal Diffusivity
Emittance	Thermal Linear Expansion
Prandtl Number	Thermal Volumetric Expansion
Reflectance	Transmittance
Specific Heat	Viscosity
Electronic, Electrical, Dielectric, Optical and Magnetic Properties	
Absorption Coefficient	Luminescence properties
Dielectric Constant	Magnetic Hysteresis
Dielectric Strength	Magnetic Susceptibility
Effective Mass	Magnetoelectric Properties
Electrical Hysteresis	Magnetomechanical Properties
Electrical Resistivity	Mobility
Electron Emission Properties	Photoelectronic Properties
Energy Bands	Piezoelectric Properties
Energy Gap	Refractive Index
Energy Levels	Thermoelectric Properties
Hall Coefficient	Work Function

computerized, and is currently being extended to give it on-line interactive capability. The coverage includes metallic and nonmetallic elements, organic and inorganic compounds, ferrous and nonferrous alloys, intermetallics, ceramics, cermets, glasses, composites, coatings and polymers. It extends also to materials beyond the immediate concern of nuclear engineering, such as foods and various biologicals.

The Materials Properties Numeric Data System is fully computerized, with on-line interactive capability planned for the near future. Due to budgetary constraints, so far only about 10%, or 10,000 data sets out of a total of approximately 100,000, have been transferred to the magnetic medium. Computerization of the collections of the Center is organized in such a manner that both the numerical data and the associated bibliographic information on its sources are retrieved simultaneously.

In addition to bibliographic and data search and retrieval from its data bases to meet the needs of its users, CINDAS offers to its users other services, such as data evaluation, data estimation and advisory and consulting services. The charges for these services are about $40 per manhour. Judging by past experience, a typical charge incurred in response to a request for data evaluation or prediction may vary from $400 to $1,000 for a single property of a single material. The charge for a single bibliographic search is currently $40.

REFERENCES

1. Y. S. Touloukian, ed., "Thermophysical Properties Research Center Data Book," 3 Vols., Purdue University, Thermophysical Properties Research Center, 3322 pp. (1960–66).
2. Y. S. Touloukian, *Thermophysical Properties of High Temperature Solid Materials*, 6 Vols. (9 books), Macmillan Co., New York, 8615 pp. (1967).
3. Y. S. Touloukian and C. Y. Ho, eds., *Thermophysical Properties of Matter—The TPRC Data Series*, 14 Vols. (15 books), IFI/Plenum Data Co., New York, 16,810 pp. (1970–79).
4. Y. S. Touloukian and C. Y. Ho, eds., *McGraw-Hill/CINDAS Data Series on Material Properties*, McGraw-Hill Book Co., New York (1981–). (Projected to comprise a total of 42 volumes and 15,000 pages.)

DOD Mechanical Properties Data Centers

Two of the DOD Information Analysis Centers containing data of interest to the nuclear industry are located at the Battelle Columbus Laboratories, 505 King Avenue, Columbus, OH 43201, and operate under the direction of Dr. Harold Midlin. They are the Mechanical Properties Data Center and The Metals and Ceramics Information Center.

Collections of the Mechanical Properties Data Center include, among others, data on: low alloy steels, stainless steels, nickel and chromium base alloys, plutonium and uranium alloys, beryllium, zirconium and hafnium metals, as well as carbon. A computerized storage, retrieval and display system of the Center carries data from conventional mechanical property tests. Both tabular and graphical representations of the data content can be handled. Computerized data analysis is possible through the use of suitable routines. The properties obtained from standard tests include: tension, compression and shear moduli, vessel burst data, creep, stress corrosion, fatigue crack propagation and fracture toughness data. High and low cycle fatigue and impact strength are also included.

The Metals and Ceramics Information Center (MCIC), is an outgrowth of the Ceramics and Graphite Technical Evaluation Center, originally located at the Wright-Patterson Air Force Base, and the Titanium Metallurgy Laboratory of Battelle. After several changes of name, the two bodies merged in 1971 to form the present entity.

The Center collects information on production processes by forging, rolling and extrusion, joining, surface treatment and powder technology, as applied to metals and ceramics. Among the metals included in the data base of the center are: aluminum, beryllium, magnesium, refractory metals, titanium, high-strength steels and superalloys. Among the ceramics covered are: borides, carbides, nitrides, oxides, sulfides and silicides, glass and glass ceramics, as well as graphite.

MCIC makes use of the bibliographic data base of the Battelle Columbus Laboratories which at the present time contains over 130,000 document and journal references. The Center collects data and publishes information on corrosion, mechanical properties, fracture toughness and processing of metals and ceramics. Information published by the Center on select cryogenic properties of materials[(1),(2)] is of particular interest to the development of fusion reactor technology, with reference to superconducting magnetic coils and neutral beam injectors. Also of interest to nuclear applications are the Center's handbooks on metallurgical treatment of uranium,[(3),(4)] and of beryllium.[(5),(6)]

Other publications by MCIC cover corrosion and fracture toughness of materials. Titanium, superalloys, ceramics and composites are covered by reports and handbooks issued by the Center.

REFERENCES

1. "Low Temperature Properties of Selected Materials—A Bibliography with Descriptors, 1948–1976," Metals and Ceramics Information Center, Columbus, OH (1977).
2. "Handbook on Materials for Superconducting Machinery," Metals and Ceramics Information Center, Columbus, OH (1974), Suppl. 1 (1975) Suppl. 2 (1977).
3. "Processing and Applications of Depleted Uranium Alloy Products," Metals and Ceramics Information Center, Columbus, OH (1976).
4. "Physical Metallurgy of Uranium Alloys," *Proc.* 3rd Army Materials Techn. Conf., Metals and Ceramics Information Center, Columbus, OH (1974).
5. "Nontraditional Machining of Beryllium," Metals and Ceramics Information Center, Columbus, OH (1972).
6. "Nondestructive Testing of Beryllium," Metals and Ceramics Information Center, Columbus OH (1973).

Materials for Nuclear Space Reactors

Application of nuclear reactors in space, through either the SNAP program or the NERVA program, has contributed to the development of data bases for materials properties. The first such data base was developed by the Southwest Research Institute in San Antonio, TX, for energy conversion materials.[(1)] The materials covered included liquid metals: mercury, cesium, rubidium, potassium, NaK(78), sodium, lithium, bismuth and lead; nonmetals: aluminum bromide, sulfur and lithium hydride; gases: argon, helium and hydrogen.

For pressures from 1 to 20 atmospheres, and temperatures up to 2300°F for mercury and 4500°F for lead, vapor pressure, density, viscosity, sur-

Table 2. Materials in the NERVA Data Book

Vol.	Materials Covered
1	Light Metal Alloys of:
	Aluminum
	Titanium
1A	Nickel Base Alloys:
	Solution Hardening
	Precipitation Hardening
	Dispersion Hardening
	Refractory Metals and Alloys:
	Molybdenum
	Titanium
	Tungsten
	Other Nonferrous Metals and Alloys:
	Cobalt
	Copper
	Zirconium
2	Ferrous Alloys:
	Carbon and Low Alloy Steels
	Austenitic Stainless Steels
	Martensitic Stainless Steels
	Precipitation-Hardening Stainless
	Iron–Nickel–Chromium
	Bolt Materials
3	Nuclear Materials:
	Nuclear Fuel Materials
	Neutron Control Materials
	Moderator Materials
	Graphites:
	Structural Graphites
	High-Expansion Structural Graphites
	Pyrolytic Graphite
	Other Graphites
	Refractory Ceramics:
	Oxides
	Carbides
	Nitrides
	Adhesives:
	Epoxies
	Phenolics
	Polyurethanes
	Silicones
	Ceramics (Cements)
	Plastics

face tension, electrical resistivity, thermal conductivity, specific heat, latent heats, enthalpy–entropy relationships, melting point, critical properties, dielectric constant, ionization potential, magnetic susceptibility, thermal neutron cross sections and corrosion properties are presented in graphs and tables.

The development of the Nuclear Engine for Rocket Vehicular Applications (NERVA), created a need to collect and evaluate properties of materials retaining mechanical strength at extreme temperatures and in high radiation fields. The unusually wide range of operating temperatures was predicated by the design of the engine, where the liquid hydrogen propellant fluid entered the beryllium reflector of the NERVA reactor at cryogenic temperatures, and exited the rocket nozzle at 4200°R. In response to this need, the Aerojet General Corporation assembled in cooperation with the Westinghouse Astronuclear Laboratory a materials prop-

Table 3. Material Properties Included in the NERVA Data Book

General description of material
Commercial designations
Form in which available
Chemical composition
Room temperature mechanical properties:
Ultimate tensile strength
Tensile yield strength
Compressive yield strength
Ultimate shear strength
Ultimate bearing strength
Bearing yield strength
Elongation
Reduction in area
Modulus of elasticity
Modulus of rigidity
Compressive modulus of elasticity
Poisson's ratio
Hardness
Room temperature physical properties:
Density
Specific heat
Thermal conductivity
Mean coefficient of linear thermal expansion
Electrical resistivity
Electrical conductivity
Magnetic properties
Melting point
Boiling point

erties data base. The content of that data base was published in 1970 as a three-volume handbook.[2] Table 2 outlines the materials contained in it, and Table 3 lists their properties. Tabular and graphical presentation of data was used in the data book, with the tensile strength of each material presented graphically as a function of temperature. Graphs of fatigue strength as a function of number of cycles endured and the temperature of exposure were also given. This data base, together with formulations of the thermodynamic properties of liquid and gaseous para- and ortho-hydrogen, supplements improvements in the formulation and speed of execution of computer codes for the thermodynamic analysis of space reactors. These spinoff effects of the space program supply valuable additions to the data base and the methodology of nuclear engineering. Some of these contributions are discussed elsewhere in this guide.

REFERENCES

1. W. D. Weatherford, Jr., John C. Tyler, and P. M. Ku, "Properties of Inorganic Energy-Conversion and Heat Transfer Fluids for Space Applications," WADD TR 61-96, Southwest Research Institute, San Antonio, TX (1961).
2. "Materials Properties Data Book," 3 vols., Aerojet Nuclear Systems Company, Sacramento, CA (1970).

4.2 REACTOR AND EQUIPMENT PERFORMANCE DATA

International Reactor Performance Data Bases

The International Atomic Energy Agency in Vienna publishes a catalog: "Operating Experience With Nuclear Power Stations in Member States," IAEA-127. The coverage extends over all known operating power and experimental reactors, but the coverage of the data given is quite uneven, with very incomplete information on reactors built in communist-controlled countries, due to their reluctance to disclose failures. For other countries the disclosure record is better, but even there a recent Swedish study[1] of a computerized data base for the failure rate of plant components concluded that it is only 50% complete, and that the coverage and description of common cause failures and operator errors are insufficient for the needs of probabilistic risk assessment analyses.

The British journal *Nuclear Engineering International* publishes annually a useful catalog of existing reactors, their design data and operating status. It appears as a supplement to the July/August issue.

On off-performance data, the Canadian Chalk River Nuclear Laboratories have been publishing annually "Steam Generator Tube Failures:

World Experience in Water-Cooled Reactors," which contains much useful data in this area.[(2)–(4)]

A good part of the material contained in the above data collections has not yet been placed on a magnetic medium, nor is their coverage of failure data of components in nuclear plants complete.

REFERENCES

1. J. P. Bento and E. Wendler, "Quantitative and Qualitative Analysis of the ATV Data Base," STUDSVIK/K2-81/497, Studsvik Energiteknik AB., Nykoeping (Sweden) (Mar. 1981).
2. P. D. Stevens-Guille, "Steam Generator Tube Failures: World Experience in Water-Cooled Nuclear Power Reactors During 1972," AECL-4753, Chalk River Lab., Chalk River, Ontario (1974).
3. M. G. Hare, "Steam Generator Tube Failures: World Experience in Water-Cooled Nuclear Power Reactors in 1974," AECL-5242, Chalk River Lab., Chalk River, Ontario (1975).
4. W. R. Casto, "Steam-Generator Tube Failures: World Experience in Water-Cooled Nuclear Reactors in 1975," *Nuclear Safety*, 18, No. 3, 355–364, (May–June 1977).

Domestic Equipment Performance Data Bases

From the very outset of the nuclear program it was evident to all concerned with its promotion that its merits would be judged on the basis of performance and safety. Thus, from the day the first reactor went on line, charts plotting its electrical energy output versus time appeared on the walls of engineering design offices and public utility headquarters, and in the Atomic Energy Commission. Soon these simple charts gave way to more complicated evaluations of performance, and specialized publications appeared. The numerical information stored in them constitutes a valuable data base.

One such publication containing a wealth of information on operating reactors is issued quarterly by the Nuclear Regulatory Commission under the title "Licensed Operating Reactors, Status Summary Report," NUREG-0020. Current and cumulative power generation numbers and charts are given, and operating occurrences relating to safety are listed. The summary pertains to U.S. reactors only.

Off-performance data for complete plants and their major components have been a subject of much interest, particularly in view of the recurrent problems with steam generator tube failures. Operating with water on the primary side at 15.5 MPa pressure, temperatures in the range 320–330°C, and up to 440 kW/m^2 heat fluxes, the steam generator tubes are sensitive to water chemistry. When tube wall wastage was observed and traced to the effect of phosphates used in makeup water, a switch was made in 1974 to a hydrazine or ammonia treatment.

Due to the nature of steam generators, over 50% of all impurities present in the secondary side water tend to accumulate at the base of the inverted U tubes. Even with periodic blowdowns, over 100 kg of precipitate is usually present on top of the generator tube sheet. The accumulation of crud at this location facilitates corrosion of the tubes.

Steam generator performance was the subject of a Westinghouse summary,[1] as well as a topic at several society meetings[2]–[5] and of a recently issued NRC report.[6] In all cases such problems as denting of generator tubes, the effects of combined erosion and corrosion, the intergranular corrosive attack at the grain boundaries in Inconel-600 alloy, localized pitting, and stress corrosion cracking are discussed, and the effects on both the primary and the secondary side of tubes are considered. Methods of dealing with these problems are described, including replacement and gross repair as well as tube plugging. The effect on power degradation is also considered in terms of a reduced plant rating, or decreased availability of equipment due to an increase in down time.

The establishment of another reactor performance data base is due to the efforts of Bernard J. Verna. The author of a regular feature appearing every other month in *Nuclear News*, for years he focused attention on problems occurring in nuclear power plants, and he described methods of dealing with them. He has made this information available in more depth in the pages of "Nuclear Power Experience," an information service now published by the Petroleum Information Co., P. O. Box 2612, Denver, CO 80201. The information on operating experience collected by Verna over the years has grown into a respectable data base. It is available as a five volume set of loose-leaf reference data:

- BWR-1: Contains brief descriptions of BWR plants, issued at about the time of plant start-up, and their subsequent operating histories. Indexed by plant name.
- BWR-2: Accumulates the current operating problems of each BWR plant, and is indexed by plant system and component, including plant cross references.
- PWR-1: Consists of brief description of PWR plants at startup, as for BWR.
- PWR-2: Accumulates operating problems of PWR plants, as above.
- Index volume and disseminating instructions by department.

The data condensed in these volumes is obtained by regular scanning of some 20 publications and serial reports issued in the field. Complete five-volume sets containing all the information collected to date are available to domestic subscribers at $2000 a set, and the annual updates sent on a monthly basis cost just over $1000 a year.

REFERENCES

1. "Steam Generator Report on Design, Operations and Chemistry Requirements," Westinghouse Water Reactors Divisions, Pittsburgh PA (Feb. 1975).
2. W. H. Layman et al., "Status of Steam Generators," *Proc.* of the Amer. Power Conf., 41, p. 65 (1979).
3. S. J. Green and D. A. Steininger, "Pressurized Water Reactor Steam Generators—Problems and Progress," ASME Nucl. Eng. Conf. San Francisco (Aug. 19–21, 1980).
4. C. L. Williams and S. J. Green, "Thermal and Hydraulic Aspects of PWR Steam Generators," ANS/ASME Int. Topical Meeting on Nuclear Reactor Thermal Hydraulics, Saratoga Springs, NY (Oct. 1980).
5. J. A. Mundis, "Challenges in PWR Steam Generators," *Chemical Engineering Progress*, 79, 39–46 (July 1983).
6. "Steam Generator Tube Experience," NUREG-0886, Nuclear Regulatory Commission (1982).

In-Plant Reliability Data Bases

A major domestic program concerned with the performance of reactor components has been under way since 1978, sponsored by the American National Standards Institute through its Failure and Incidents Reports Review. The objective of this program is to collect data on the component population and equipment maintenance records from nuclear power stations, and to incorporate the resulting data in the In-Plant Reliability Data (IPRD) file. Data encoding, collection and analysis are directed by the Oak Ridge National Laboratory, and are partially funded by the Office of Nuclear Regulatory Research of the U.S. Regulatory Commission.

The three parts of the data base are: the population record, the failure record and the repair record. The first of these contains information about the design, operating mode and environment plus, the functional name of the component. All components are entered into the record without regard to their failure history or functional importance to safety. The second part, the failure record, is constructed on the basis of failure description by analysts, who assign to each entry the failure mode, cause and severity level: incipient, degraded and catastrophic. Finally, the repair record contains repair time, category and description. Each maintenance action results in the creation of one repair record.

Thus far 10 nuclear power stations, operating a total of 16 light water reactors, have been visited and have provided data on their pump population, and failure and repair information. After review for completeness, some 120,000 maintenance records, originating from six reactors operated at four plants, were incorporated in the data base. Using as a base 1468 pumps located at these plants, 3100 maintenance records, covering some 24 reactor years of commercial operation at power, plus 900 maintenance records obtained prior to commercial operation, were included in this data base.

This data base incorporates a set of generic system definitions to allow comparison of components performing the same function but possibly given different names in different plants. An overview of the program plus the generic definitions of the components are contained in the methodology report for the program.[(1)] The component performance data base thus contained in IPRD is remarkable for its completeness in reporting both the population and the frequency of component failures. Further extension of the data base to more plants is planned in the future, together with periodic updating of the existing records.

REFERENCES

1. J. P. Drago, R. J. Borkowski, D. H. Pike, and F. F. Goldberg, "The In-Plant Reliability Data Base for Nuclear Power Plant Components: Data Collection and Methodology Report," NUREG/CR-2641 (ORNL/TM- 8271), Oak Ridge National Lab. (July 1982).

4.3 REACTOR SAFETY DATA BASES

The data bases storing information relevant to reactor safety encompass many diverse fields. The response of reactor equipment to operational transients, and of the reactor vessel materials to radiation damage, embrittlement and thermal shock are of primary interest in this discipline.

Of particular interest here are the changes in the properties of materials on irradiation, with attention centered on the increase of pressure vessel brittleness, and its possible susceptibility to the effects of pressurized thermal shock. Steels with higher copper, nickel and phosphorus content, used in older reactor vessels, require particular attention.

Safety studies use properties of materials for core damage calculations in water and liquid metal cooled reactors, as well as the seismic data for the evaluation of reactor safety in case of an earthquake.

Equipment Transients Data Base

The main data base of the Electric Power Research Institute containing results from transient tests of nuclear reactor equipment is supported by the DATATRAN[(1)] executive code system. The system is principally an extension of the FORTRAN programming scheme, and is designed to handle the functions of data management.

Originally developed in the late 1960s at Knolls Atomic Power Laboratory in Schenectady, NY, as a general data base management system, DATATRAN has been lately adapted under the auspices of the Electric Power Research Institute, at the cost of some 40 man-years of effort, to handle the large amount of data obtained from transient tests of reactor equipment. In addition to storing the data, it also implements the input, output and control functions. The code system is very easy to use in spite of its large size and complex functions.

In its application,[(2)] the system is used for inserting lines of DATATRAN coding into existing FORTRAN codes. Alternately, DATATRAN modules, similar to FORTRAN subroutines, can be written to specifications and inserted into existing programs to perform such data manipulations as required. Any code thus modified will be characterized by a very high data transfer rate and access to all available core memory. DATATRAN interfaces with the EPRI Reactor Analysis Support Package (RASP)[(3)] used for core reload analyses. Operation of RASP involves the use of other computer codes such as RETRAN[(4)] and FRAY.[(5)]

The principal data sets included in the system, and available to the nuclear industry through EPRI, are listed in Table 4. The intent is to include in this data bank all available transient data from startup tests, tests during rise to full power, plus accidental and off-normal transients

Table 4. Content of EPRI DATATRAN Transient Tests Data Bank

Test Description	No. of Data Entries
CE/EPRI 1/5 Scale Pump Performance Tests	
962 Steady State Tests	190,000
14 Transient Tests	15,500,000
B & W 1/3 Scale Pump Performance Tests	
860 Steady State Tests	51,000
Semiscale Pump Performance Tests	
354 Steady State Tests	4,800
Marviken Critical Flow Studies	
27 Transient Tests	15,000,000
Three Mile Island, Unit-2 Accident	
1 Transient	500,000
Peach Bottom Turbine Trip Tests	
3 Transient Tests	4,500,000
Arkansas Nuclear One, Unit-2	
11 Transient Tests	350,000

for commercial nuclear plants. The potential uses of this system are now being defined for electric utilities operating nuclear reactors. Its use as a general data management system, and as a general computer code executive system, is being promoted through seminars, formal workshops and informal tutorials.

DATATRAN has also been chosen by the U.S. Nuclear Regulatory Commission to handle the Nuclear Plant Parameters Data Bank. This data bank currently contains the physical description of Zion-1 PWR plant, with the plant geometry and its operating parameters described by some 150,000 data entries, each recorded and retrievable through a hierarchical interrogative data structure.

REFERENCES

1. "Organization and Structure of an EPRI DATATRAN Data Bank," Informal Report, Electric Power Research Institute, Intermountain Technologies Inc. (June 1980).
2. H. Kopp, "DATATRAN User's Guide," KAPL-M-7372, Rev. 1, Knolls Atomic Power Lab. (Aug. 1977).
3. L. Agee et al., "RASP: An Integrated Reactor Analysis Support Package," Informal Report, Electric Power Research Institute (June 1982).
4. K. V. Moore et al., "RETRAN—A Program for One-Dimensional Transient Thermal-Hydraulic Analysis of Complex Fluid Systems," EPRI CCM-5, Electric Power Research Institute (Dec. 1978).
5. Y. R. Rashid and R. S. Dunham, "FRAY—A Program for Two-Dimensional Transient Fuel Rod Analysis, Vol. 1: Theoretical and Numerical Bases," ANA82-1, and "Volume 2: User's Manual," ANA82-2, ANATECH (Jan. 1982).

RVS—Data Base for PWR Vessel Surveillance

This data base was developed by EPRI to provide a comprehensive computerized data base for reactor vessel surveillance program. It arose out of two earlier EPRI data bases: Nuclear Pressure Vessel Steel Data Base,[(1)] established in 1978, and the Irradiated Nuclear Pressure Vessel Steel Data Base,[(2)] published in 1982. It represents data from 146 heats of steel used to fabricate 50 reactor vessels. In addition, data from 73 surveillance capsules is included. These surveillance capsules are mounted at various points in the interior of the reactor pressure vessel, and are exposed to the same conditions of irradiation and water chemistry as the reactor vessel itself.

Data for each heat of steel includes heat chemistry, irradiation parameters, and mechanical property test results such as fracture toughness, Charpy V-notch test, tensile test results, nil-ductility transition temperature. Statistical analyses of the Charpy V-notch test include specific temperature-energy ranges of interest, hyperbolic tangent curve fit parameters, and lower bound of upper shelf energy.

EPRI is currently sponsoring the development of a comprehensive data base for all information relating to reactor pressure vessels in a broad base attack on the pressurized thermal shock problem.[3]

REFERENCES

1. T. U. Marston, W. L. Server, and W. Oldfield, "Nuclear Pressure Vessel Steel Data Base," Topical Report NP-933, Electric Power Research Institute (Dec. 1978).
2. T. U. Marston, P. McConnell, W. L. Server, W. Oldfield, and F. M. Oldfield, "Irradiated Nuclear Pressure Vessel Steel Data Base," Topical Report NP-2428, Electric Power Research Institute (Jun. 1982).
3. Theodore A. Meyer, Kenneth R. Balkey, and Stephen E. Yanichko, "The Evolution of Reactor Vessel PTS—A Catalyst to the Advancement of Technology," *Mechanical Engineering*, 106, 34–46 (June 1984).

MATSURV—Data Base for Pressure Vessel Material Surveillance

This data base was established by NRC as part of the effort to collect and systematize information derived from the licensee surveillance programs for reactor pressure vessels.[1] Initially developed under a contract given by Department of Energy to the Sandia National Laboratory, Albuquerque, NM and subcontracted to the International Energy Associates Limited, the data base became available in 1980. The data base is operational on the IBM 370/168 computer, located at the National Institutes of Health, with access limited to government agencies. Data retrieval from MATSURV and output formatting utilizes WYLBUR online interactive text editor.

The material contained in this data base was initially obtained from responses to an NRC questionnaire, distributed to licensees in 1977. Data from five reactor vendors covering 65 operating reactors was included in the initial data entry. After the data base became operational in 1980, surveillance capsule reports obtained by Westinghouse up to March 1981 were incorporated.

The materials data contained in MATSURV is in three broad categories, arranged in a five-level hierarchial structure. The categories are: (a) reactor pressure vessel description, (b) preirradiation surveillance baseline, (c) irradiated surveillance program data. The five hierarchial levels are: (1) general information based on reactor docket number, (2) identifiers for basic information in three material categories (vessel, baseline and irradiated), and identifiers for heat treatment sequences plus chemical compositions, (3) detailed information on material sets, heat treatment temperatures and hold times, (4) test results from the baseline specimens, (5) test results from the irradiated surveillance specimens.

All information contained in the data base is identified in the system users manual[2], and filed under the reactor docket number. The data base concentrates on the properties of the materials used in the fabrication of the plates constituting the reactor pressure vessel beltline region. This is the region of highest radiation damage and concentration of thermal stresses.

REFERENCES

1. A. F. Billy and T. U. Marston, "Role of the Computerized Material Surveillance Information System on Reactor Vessel Integrity Assessment," ASME/ANS Second Joint Nuclear Engineering Conference, Portland, OR (July 26, 1982).
2. J. Strosnider, C. Monserrate, L. D. Kenworthy, and C. D. Tether, "Computerized Reactor Pressure Vessel Materials Information System," NUREG-0688, U.S. Nuclear Regulatory Commission (Oct. 1980).

MATPRO—Data Base for Core Damage Calculations

Currently a concerted effort is being made to supplement and systematize the collections of descriptions of materials properties relevant to reactor safety, and to place them within existing data bases. Where appropriate data bases do not exist, new bases are created. One such effort is being sponsored by the U.S. Nuclear Regulatory Commission, and involves the development of a handbook of models for LWR material properties relevant to core damage analysis.[1] It bears the name MATPRO,[2] and is intended for use with computer codes analyzing the behavior of LWR fuel rods, such as FRAPCON[3] and FRAP-T.[4] Other fuel performance codes can use the data contained in MATPRO due to its generalized format.

At present MATPRO contains more than one hundred computer models dealing with light water reactor fuels, fuel rod cladding, control rod cladding, neutron absorbers, grid spacers, materials properties of gas-gap mixtures, and fuel cladding–oxygen compounds. The models are either polynomial expressions with constants obtained from a least squares fit to existing available data, or analytical expressions of a form suggested by theory but with constants determined from experimental data. As a rule, the primary variable in the models is the temperature; some models also include burnup, time and other variables. Table 5 lists the primary properties considered for the six materials most relevant to fuel design: mixed U–Pu oxides, Zircaloy, B_4C, Zr–U–O mixes, stainless steel 304 and gases of interest.

Table 5. Materials Properties Contained in MATPRO

Property Mat.:	(U,Pu)O_2	Zircaloy	B_4C	Zr–U–O	SS-304	Gases
Specific heat	x	x	x	x	x	—
Enthalpy	—	x	x	x	x	—
Thermal conductivity	x	x	x	x	x	x
Emissivity	x	x	—	—	—	—
Thermal expansion	x	x	x	x	x	—
Elastic modulus	x	x	—	—	—	—
Creep rate	x	x	—	—	—	—
Melting temperature	x	x	x	x	x	—
Oxidation	x	x	—	—	x	—
Viscosity	x	—	x	x	—	x
Densification	x	—	—	—	—	—
Swelling	x	—	—	—	—	—
Pressure sintering	x	—	—	—	—	—
Restructuring	x	—	—	—	—	—
Fraction strength	x	—	—	—	—	—
Fission gas release	x	—	—	—	—	—
Cs and I release	x	—	—	—	—	—
Vapor pressure	x	—	—	—	—	—
Axial growth	—	x	—	—	—	—
Plastic deformation	—	x	—	—	—	—
Annealing	—	x	—	—	—	—
Texture factors	—	x	—	—	—	—
Embrittlement	—	x	—	—	—	—
Cyclic fatigue	—	x	—	—	—	—
Collapse pressure	—	x	—	—	—	—
Hydrogen uptake	—	x	—	—	—	—
Meyer hardness	—	x	—	—	—	—
Coeff. of friction	—	—	—	x	—	—
Surface tension	—	—	—	x	—	—
UO_2–Zr react. rate	—	—	—	x	—	—
UO_2 solut. heat	—	—	—	x	—	—

x = property included in MATPRO.

REFERENCES

1. D. L. Hagrman, "Materials Properties Models for Severe Core Damage Analysis," EGG-CDD-5801, EG&G Idaho, Inc. (May 1982).
2. D. L. Hagrman et al., "MATPRO-Version 11 (Revision 2): A Handbook of Materials Properties for Use in the Analysis of Light Water Fuel Rod Behavior," NUREG/CR-0479, TREE-1280, Rev. 2 (Aug. 1981).
3. G. A. Berna et al., "FRAPCON-2: A Computer Code for the Calculation of Steady-State Thermal-Mechanical Behavior of Oxide Fuel Rods," NUREG/CR-1845, EG&G Idaho, Inc. (Jan. 1981).
4. L. J. Siefken et al., "FRAP-T6: A Computer Code for the Transient Analysis of Oxide Fuel Rods," EGG-2104, NUREG/CR-2148, EG&G Idaho, Inc. (1981).

SACRD—Data Base for LMFBR Safety Calculations

The Oak Ridge National Laboratory has been engaged since 1976 in the collection and systematization of data for use in computer codes applied to safety research on fast reactors. A data base named SACRD[1] (Safety Analysis Computerized Reactor Data) was created and placed on a magnetic medium for on-line access. The data base is maintained by means of the ORNL data management system JOSHUA.[2]

The effort connected with the operation of this data base is directed by the Safety Analysis Data Coordination Group (SADCG), formed by experts in fast reactor work from industry, national laboratories, government and universities. SADCG coordinates the activities of subcommittees charged with the task of evaluating the data to be entered into SACRD. The first version of SACRD was released in 1978, and subsequently revised. The latest version was released in October 1981.[3]

The data base contains nuclear, thermophysical, structural, aerosol and radiological data for more than 70 materials, from single isotopes through mixtures, and composites such as concrete. More than 150 properties, such as heat capacity, thermal conductivity and so on, are included, and thermophysical properties of fuel ceramics such as UO_2, PuO_2 and $(U,Pu)O_2$ are of particular interest.

The presentation of these properties in the data base is very thorough, with abstracts describing the sources of evaluations and the theoretical models used. Data are entered either in tabular form or as fits into appropriate FORTRAN subroutines. Uncertainty information about the data is entered into separate files.

Remote access to SACRD is possible through on-line facilities on approval from the Director of the Office of Reactor Research and Technology, DOE, and is limited to U.S. citizens actively engaged in development of breeder reactors. An update on the services currently available from SACRD is maintained through *SACRD Bulletin*, mailed to personnel engaged in U.S. breeder reactor safety analysis.

REFERENCES

1. N. M. Greene, V. M. Fosberg, G. B. Raiford, J. A. Arwood, D. B. Simpson, and G. F. Flanagan, "SACRD: A Data Base for Fast Reactor Safety Computer Codes," Oak Ridge National Laboratory (Jan. 1979): "General Description," ORNL-5477/V1, and "Contents and Glossary of Version 1 of the System," ORNL-5477/V2.
2. H. C. Honeck, "The JOSHUA System," DP-1380 (Apr. 1975).
3. N. M. Greene, V. M. Fosberg, G. B. Raiford, J. W. Arwood, J. F. Manneschmidt, and G. F. Flanagan, "SACRD: A Data Base for Fast Reactor Safety Computer Codes—Version 81," ORNL/CSD-86, Oak Ridge National Lab. (May 1982).

Seismic Safety Data Collections

Potential markets for nuclear reactors are, as a rule, located away from fossil fuel deposits. In such locations fuel costs for power generating stations are strongly influenced by fuel transportation costs. One location with high fossil fuel costs but a high economic incentive for building nuclear power plants as well, is the west coast of the continental United States. The same area is also subjcct to frequent earthquakes; so almost from the inception of the nuclear industry, attention has been focused on assuring earthquake safety to nuclear power stations located in areas subject to seismic hazards.[(1),(2)]

The design methodology of nuclear plants for earthquake protection was formalized early by Hansen,[(3)] and regulatory efforts of the U.S. government followed. In Hansen's work, accumulated knowledge on the subject of earthquakes and methods for preventing personal, equipment and plant damage was of immense value.

Problems connected with earthquake safety assurance fall into several categories: identification of hazard areas, selection of plant sites with the lowest risk within a given area, assurance of operability of plant control equipment under conditions of seismic hazard, and protection of major and minor parts of the plant installation from damage in the event of a quake.

The identification of hazard areas is much facilitated by use of data on location and frequency of earthquakes, available from the National Geophysical and Solar Data Center at Boulder, CO,[(4)] and the data bases of the California Institute of Technology and the University of California at Berkeley. Selection of plant sites with the least seismic risk is governed by the NRC site selection criteria,[(5)] as formulated by Regulatory Guide 4.7.

Reduction of hazards due to the inoperability of plant control instrumentation is achieved by designing and testing plant control panels and the associated equipment to withstand the vibration environment encountered during an earthquake.[(6)] Vibration amplitudes and frequencies recorded during actual earthquakes[(7)] are reproduced in tests, using equipment developed during the space program for vibration testing of rocket engines. This equipment is powerful enough to accommodate full-wall control panels of a power plant.

Inoperability of the rest of the plant during an earthquake is minimized, and the threat to its integrity eliminated, by providing all important pieces of cquipment with yielding constraints, and by placing an adequate number of hydraulic snubbers to meet the requirements of damping protection.[(8)]

Seismic risk analysis and the seismic fragility of nuclear plant components have been the object of separate studies.[(9),(10)]

REFERENCES

1. E. D. Bloom and R. C. Erdman, "Prediction of Earthquake Magnitude Versus Likelihood of Selected Surface Regions of the Earth," *Trans. ANS,* 27, 633–634 (Nov. 1977).
2. T. F. Lomenick, ed., "Earthquakes and Nuclear Power Plant Design," ORNL-NSIC-28, Nuclear Safety Information Center, Oak Ridge, TN (1970).
3. Robert J. Hansen, *Seismic Design for Nuclear Power Plants,* The MIT Press, Cambridge, MA (1970).
4. "A Computer Type of World Earthquake Events," National Geophysical and Solar Data Center, Boulder, CO (1977).
5. "General Site Suitability Criteria for Nuclear Power Stations," Regulatory Guide 4.7, Nuclear Regulatory Commission, Washington, DC (1975).
6. "Seismic Design Classification," Regulatory Guide 1.29, Nuclear Regulatory Commission, Washington, DC (1978).
7. "Design Response Spectra for Seismic Design of Nuclear Power Plants," Regulatory Guide 1.60, Nuclear Regulatory Commission, Washington, DC (1973).
8. "Damping Values for Seismic Design of Nuclear Power Plants," Regulatory Guide 1.61, Nuclear Regulatory Commission, Washington, DC (1973).
9. P. D. Smith, D. L. Bernreuter, et al., "An Overview of Seismic Risk Analysis for Nuclear Power Plants," Structural Mechanics Associates (Sept. 1980).
10. Robert P. Kennedy, R. D. Campbell, et al., "Subsystem Fragility," Seismic Safety Margin Research Program, NUREG/CR-2405 (Feb. 1982).

4.4 DATA ON SPENT FUEL AND WASTE DISPOSAL

Integrated Data Base for Radioactive Waste

The radioactive waste data base was established at the Oak Ridge National Laboratory under the sponsorship of the Department of Energy to maintain a current, and as complete as possible, inventory of spent reactor fuel, and of other radioactive waste. Sources of radioactive waste from both commercial and government installations are considered, the aim being to provide not only historical and current inventories but also projections for the future based on the current reactor operating records.

For purposes of classification, radioactive waste is divided into three categories:[(1)] (a) high level waste, (b) transuranic waste, (c) low level waste.

Category (a) comprises inventories of fission products from operating reactors mixed with unused reactor fuel.

Category (b) includes inventories of elements artificially produced in fission reactors, such as plutonium, neptunium, einsteinium, curium, americium, californium and berkelium.

Category (c) lumps together all other radioactive materials produced by reactor irradiation or generated in various medical or cleanup processes.

Within each classification the waste is characterized by its volume, in cubic meters; radioactivity, in curies (1 Ci = 3.7×10^{10} dis/sec); thermal power, in kilowatts; and mass, in kilograms of transuranic elements. Since radioactivity is time dependent, and as a rule decreases with time, all radioactive inventories are adjusted for time of decay. This adjustment is done by means of the ORIGEN code.[(2),(3)]

The sources of radioactive waste entered in this data base include uranium mining and milling, conversion to the hexafluoride form and enrichment, fuel fabrication and reprocessing, medical uses, military naval reactor wastes, research and developmental activities. The current waste inventory and its projections are published as of the end of each year.[(4),(5)]

To achieve total accounting of waste from the flow stream of a reprocessing plant, the stream is split into all components, be they solids or liquids, and a fraction of each element in the spent fuel is assigned to each waste stream. Appropriate radioactive decay corrections are made to account for the time factor. Current information contained in this data base is collected from various sources, verified and entered into the data base in a proper format. In the process, the values entered are reconciled to form a consistent set. In handling waste inventory projections, data are entered into the data base as if the future wastes were past inventories.

In its present form, the data base provides information in an integrated and internally consistent form on the past, present and future inventories of spent fuel and wastes. It is capable of generating printed reports in various formats, and for different categories of waste. Thus it constitutes a source of data for policy analysis, long-range planning and public information. Information on this data base is available from K. J. Notz or Charles W. Fosberg, ORNL, P. O. Box X, Oak Ridge, TN 37830, tel. (615) 574-6632.

For public utilities operating nuclear plants and preparing to join one of the regional waste repository compacts prescribed by recent legislation,[(6)] this data base supplies information on total spent fuel inventories in their region. Given projections of their own and neighboring states' inventories of spent reactor fuel, plus estimates of storage and transportation costs, the utilities are in a much better position to make informed decisions regarding the siting of proposed away-from-reactor (AFR) repositories permitted by the legislation.

The Nuclear Regulatory Commission participates in these activities through its licensing involvement. In 1981 it commissioned the preparation of a review of all activities concerned with radioactive waste management.[(7)]

REFERENCES

1. "Waste Management and Disposal," Report of INFCE Working Group 7, International Atomic Energy Agency, Vienna (1980).
2. A. G. Croff, "ORIGEN2—A Revised and Updated Version of the Oak Ridge Isotope Generation and Depletion Code," ORNL-5662, Oak Ridge National Lab. (July 1980).
3. A. G. Croff, "A User's Manual for the ORIGEN2 Computer Code," ORNL/TM-7175, Oak Ridge National Lab. (July 1980).
4. "Spent Fuel and Radioactive Waste Inventories and Projections as of December 31, 1980," DOE/NE-0017, U.S. Department of Energy (Sept. 1981).
5. "Spent Fuel and Radioactive Waste Inventories and Projections as of December 31, 1981," DOE/NE-0017/V2, U.S. Department of Energy (Sept. 1982).
6. "Nuclear Waste Policy Act of 1982," H.R. 3809 (Jan. 7, 1983).
7. J. J. Clancy, O. I. Oztunali, C. J. Pitt. E. D. Picazo, et al., "Data Base for Radioactive Waste Management," NUREG/CR-1759: Vol. 1, "Review of Low-Level Radioactive Waste Disposal History"; Vol. 2, "Waste Source Options Report"; Vol. 3, "Impacts Analyses Methodology Report"; prepared for NRC by Dames and More, Inc. (Nov. 1981).

ORNL–ICRP Dosimetric Data Base

This data base is an outgrowth of cooperation between Oak Ridge National Laboratory and the International Commission on Radiological Protection. The latter body is concerned with setting limits on exposure permissible to the general public, as well as to people working professionally with sources of radioactivity, handling radioactive materials or exposed to the possibility of ingesting them either through the respiratory or the digestive tract.

The task of setting permissible exposure limits requires calculations involving 94 elements represented by 756 radionuclides. The data required include the decay schemes and half lives, number of nuclear transformations in each decay chain, specific effective energies of radiation emitted, metabolic data for each chemical element, annual limits of contaminant intake, committed dose equivalents and other similar factors.

The task of supplying and managing the above data is particularly difficult and extensive for isotopes with half-lives greater than 10 minutes, due to the prevalence of long lived isotopes among the contaminants. The establishment of the ORNL data base for dosimetry is intended to ease the large data management problem.

Data types contained in the ORNL-ICRP data base are given in Table 6. They correspond to data specified in Publications 2, 26 and 30 of ICRP.[(1)–(5)] The information is computerized, and accessible for remote on-line searching by means of the ORLOOK program, implemented on the IBM-3033 computer.

The current concern with exposures of large populations to low level radiation has focused the attention of the health physics community on

Table 6. Unabridged Numerical Data in ORNL-ICRP Data Base

Radioactive Decay Data from ENSDF	Decay Scheme Plot Files
Tabular Summary of ENSDF Data	Specific Effective Energies
Radiation Energies, Intensities	Metabolic Data
Beta Spectrum Data	Transformations/Unit Intake
Bremsstrahlung Data	Committed Dose Eq./Unit Intake
Mean Energies and Intensities	Annual Limit on Intake and Derived Air Concentration

methods of measurement and detection of man-made radiation at concentrations below 1 nCi/g, appearing against a background of comparable natural radiation. In these circumstances, prevention of sample contamination and enhancement of sensitivity of the instrumentation, as well as the selection of proper resolution vis-a-vis the efficiency of the detection apparatus, are questions of importance. Publication 22 of the International Commission on Radiation Units[6] addresses these issues and discusses the techniques to be used in measurements.

REFERENCES

1. *ICRP Publication 2, Recommendations of the International Commission on Radiological Protection, Report of Committee 2 on Permissible Dose for Internal Radiation (1959),* Pergamon Press, Oxford (1960).
2. "ICRP Publication 26, Recommendations of the International Commission on Radiological Protection," *Annals of ICRP,* 1, 3 (1977).
3. "ICRP Publication 30, Part 1, Limits for Intake of Radionuclides by Workers," *Annals of ICRP,* 2, 3–4 (1979); Suppl. to Part 1, *Annals of ICRP,* 3, 1–4 (1979).
4. "ICRP Publication 30, Part 2, Limits for Intake of Radionuclides by Workers," *Annals of ICRP,* 4, 3–4 (1980); Suppl. to Part 2, *Annals of ICRP,* 5, 1–6 (1981).
5. "ICRP Publication 30, Part 3, Limits for Intake of Radionuclides by Workers," *Annals of ICRP,* 6, 2–3 (1981); Suppl. A to Part 3, *Annals of ICRP,* 7, 1–3 (1982); Suppl. B to Part 3, *Annals of ICRP,* 8, 1–3 (1982).
6. "Measurement of Low-Level Radioactivity," ICRU-22, International Commission on Radiation Units and Measurements, ICRU Publications, Washington, DC (1972).

4.5 GENERAL DATA BASES

Data Base of the American Chemical Society

A good example of the extension of information resources through an electronic medium is provided by the publication program of the American Chemical Society (ACS), 1155 16th Street, NW, Washington, DC 20036, tel. (202) 872-4600. Among professional societies, the ACS is the first to put full-text journals on a magnetic medium.[(1)] This pioneering effort places at the disposal of the practicing engineer the capability of direct access to original papers, including full experimental details, as they appear in the primary literature. To the extent that it includes data for uranium preparation and the chemical reprocessing of spent reactor fuel, the ACS data base is of interest to the nuclear engineering community.

Over the past two years some 500 chemists and librarians have evaluated the existing chemical data base with the objective of establishing an electronic publishing project for ACS. So far ACS has placed on-line the full text contents of 18 ACS journals covering the period from 1980 to the present, and listed in Table 7. This body of information consists of some 25,000 articles on chemical processes, representing about 115,000 pages of text. For orientation, this is equivalent in volume to three and a half times the content of the full text of the *Encyclopedia Britannica*. In 1983 an additional 40,000 pages of text from ACS journals were entered on a magnetic medium. The system contains suitable search and display features for printed text, and is continuously being upgraded with the help of ACS evaluators. As yet, no illustrations are included in the retrievable files, due, in part, to the less advanced state of standardization of electronic transmission of graphical material.

The ACS data file, containing the full text of the journals listed in Table 7, has been rented out, on a royalty basis, and resides on magnetic storage disks and computer equipment operated by the Bibliographic Retrieval Services, Inc., 1200 Route #7, Latham, NY 12110, tel. (518) 783-1161. BRS data banks, in turn, can be accessed through the TYMNET, TELENET, and UNINET networks.

For access to the ACS data file through the BRS data bank service, both 300 baud and 1200 baud (30 and 120 characters per second) transmis-

Table 7. ACS Journals Accessible for On-line Searching

Accounts of Chemical Research	*J. ACS*
Analytical Chemistry	*J. Chemical and Engineering Data*
Biochemistry	*J. Chemical Information and Computer Sciences*
Chemical Reviews	*J. Medicinal Chemistry*
Environmental Science and Technology	*J. Organic Chemistry*
I&EC Fundamentals*	*J. Physical Chemistry*
I&EC Process Design and Development	*Macromolecules*
I&EC Product Research and Development	*Organometallics*
Inorganic Chemistry	
J. Agricultural and Food Chemistry	

**Industrial and Engineering Chemistry.*

sion rates are available. Although the 1200 baud rate reduces the connect time required for the same length of text retrieved, some users consider it too high for comfortable reading; so when visual inspection of the full text is required before it is sent to the printer, the 300 baud communication rate may be preferred.

The access fees charged to the user are a sum of two components: $50 per connect hour as an ACS royalty, plus $16 to $35 per connect hour of BRS charges depending on the user's contract with BRS. UNINET, TELENET and TYMNET network lines can be used for access. The communication charges per connect hour of these networks are: $6.00 for UNINET, $7.00 for TELENET and $11.00 for TYMNET. For users who do not wish to print the material retrieved at their own terminal, but require off-line printing and mailing by BRS of the material, there is an additional fee of $0.46 per page for hard-copy printing, of which $0.28 is the ACS royalty and $0.18 is the BRS service fee.

Full text retrieval of primary literature sources offered by ACS has distinct advantages over abstract retrieval. In abstracting, only major concepts are indexed. Valuable information that is peripheral to the main theme of the paper is often missed by the abstract, but can be located with full text retrieval. Since information is added to the ACS data bank at the same time that the primary journals are printed, the data bank offers a high order of timeliness and completeness.

REFERENCES

1. *User's Manual, ACS Full-Text File Primary Journal Data Base,* American Chemical Society, Books and Journals Division, Washington, DC, May 1983.

DOE Energy Data Base

This data base is operated for the DOE by the Technical Information Center at Oak Ridge National Laboratory. It is a computer-readable comprehensive unclassified collection of bibliographic data containing indexes, citations and abstracts of the world's scientific and technical literature on energy. Among subjects of particular interest to nuclear industry, it contains information announced in the *Fusion Energy Update,* and in the current awareness bulletins: *Nuclear Fuel Cycle, Nuclear Reactor Safety* and *Radioactive Waste Management*. Information from Vol. 7 (1976) onward of "Atomindex" is also included and constitutes in effect a continuation of the discontinued "Nuclear Science Abstracts."

EDB currently contains well over 1.7 million items. During 1983 approximately 200,000 research project descriptions were added to the EDB as well as over 100,000 items of interest to the nuclear programs from pre-1976 "Nuclear Science Abstracts." The present content of the base provides descriptions of results of a worldwide investment of over $170 billion in R&D projects.[1]

The content of the data base[2],[3] is obtainable on magnetic tape from the National Technical Information Service, Springfield, VA 22161, tel. (703) 478-4805. It can be accessed directly for on-line searching of its content through the Systems Development Corporation's ORBIT system, 2500 Colorado Ave., Santa Monica, CA 90406, tel. (800) 421-7229, or through the Lockheed DIALOG system, 3215 Hanover Street, Palo Alto, CA 94304, tel. (800) 227-1960. Fees for access are to be paid directly to these organizations.

If personal searching is not possible, preordered searches are available for a fee from the Western Regional Information Center, Lawrence Berkeley Laboratory, Bldg. 50, Rm. 130, Berkeley, CA 94720, tel. (415) 843-2740.

For government installations, and government contractors, on-line access to EDB is also available through the RECON network.

REFERENCES

1. "Congressional Budget Request FY 1984," U.S. Dept. of Energy, Washington, DC, Vol. 3, p. 297 (Jan. 1983).
2. "Energy Data Base: Subject Coverage, Literature Coverage, Data Elements, and Indexing Practices," DOE/TIC-4608, Technical Information Center, Oak Ridge, TN (June 1981).
3. J. Robinson and M. Hu, "DOE's Energy Data Base (EDB) Versus Other Energy-Related Data Bases: A Comparative Analysis," LBL-12547, Lawrence Berkeley Lab., Berkeley, CA (1981).

NTIS Data Bases

By far the largest collection of data and information in the form of reports, microfilms and magnetic tapes is in the National Technical Information Service repository located in Springfield, VA. Its mission is to facilitate identification and access to information obtained through the activities of the federal government. It reproduces and distributes these materials for a fee, has annual sales of about $25 million and handles more than 25,000 items a day. A machine-readable catalog of products available from NTIS can be obtained by prospective users; a considerable part of its content is pertinent to the discipline of nuclear engineering. Hard copies of reports available for sale to the public can also be ordered on-line for mailing through DIALOG and ORBIT.

The two data bases of particular interest to nuclear engineers among the eleven distributed by NTIS are: the NTIS Bibliographic Data Base and the DOE Energy Data Base. The first is available for on-line searching on DIALOG, BRS and ORBIT systems, the second on DIALOG and ORBIT only. Both are available for batch searching through the New England Research Application Center (NERAC), tel. (203) 486-4533. These organizations remit to NTIS the annual rental fee, listed below, and collect for NTIS a usage fee, in lieu of royalty, at rates different for on-line and batch searching.

- The NTIS Bibliographic Data Base covers physical sciences, social sciences, computer science and technology, economics, business, transportation and state and local information, as well as foreign documents from the NTIS foreign exchange program. It contains over 1 million citations, with over 70,000 added annually in biweekly updates, and covers 1964 to the present. Tape containing the data base may be leased at $5,000 for the current year.
- EDB, the Energy Data Base, covers coal, petroleum, natural gas and oil energy information, fission and fusion energy generation, solar, geothermal and synthetic fuels, as well as energy conservation and storage. Bibliographic information from energy-related research in the basic sciences and engineering is also covered. Articles, books, translations and patents are included. It contains over 650,000 citations, with approximately 150,000 added annually, and covers 1967 to the present. The tape containing the data base is restricted to U.S. distribution only, and can be leased at $2,000 for the current year.

For additional information contact: David Grooms, Office of Data Base Services, National Technical Information Service, U.S. Department of Commerce, 5285 Port Royal Rd., Springfield, VA 22161, tel. (703) 487-4807.

5. MACHINE METHODS OF NUCLEAR DESIGN

5.1 CODE DEVELOPMENT AND MAINTENANCE

Development of Coding Languages

When digital computers were first used in nuclear design, all coding was done in machine language. The time required to master computer coding in machine language made this task virtually impossible for most nuclear engineers. The need arose for easing the transition between the equations used by physicists and engineers in the nuclear field, and machine instructions. Pioneering work in this area was done by Grace Murray Hopper, who in 1952 developed the first compiler designated A-0, followed in 1953 by A-2, the first compiler to handle symbolic manipulation. This laid the foundation on which IBM based its first commercial compiler for the conversion of English language instructions and mathematical equations into machine language, the perennial FORTRAN (FORmula TRANslator). The credit for inventing FORTRAN in 1954 goes to John Backus of IBM. On September 20, 1954 Harlan Herrick ran the first successful FORTRAN program at IBM. This high level language was followed by other languages, such as COBOL, PASCAL and recently BASIC, ADA and others. Although each of these languages has its preferred applications and a large circle of devotees, FORTRAN continues to retain its hold on the engineering community and is the language of preference for most nuclear and thermo-hydraulic design codes.

In 1957 Hopper completed the FLOWMATIC, the first English language data processing compiler, otherwise known as B-0, and IBM released FORTRAN-I in April 1957. It became popular, caught on and was adopted widely. Improvements and extensions of its capabilities were almost immediately suggested by its many users, and by 1958 FORTRAN-II was released. By this time, the American Standards Institute had become interested in promoting the standardization of computer programming languages and established a technical committee to produce an appropriate standard. The interaction of this committee with IBM was responsible to some extent for the fact that instead of releasing FOR-

TRAN-III, IBM decided to hold off until FORTRAN-IV was ready for release in 1966. At the same time ANSI introduced its standard ANSI X3.9-1966, commonly known as FORTRAN-66, roughly matching IBM's FORTRAN-IV in its features. In a continuing effort, the ANSI Technical Committee in 1978 produced a new standard for FORTRAN, ANSI X3.9-1978, otherwise known as FORTRAN-77. The standardization activities of the engineering societies in the computer area are discussed more fully in a later chapter. Both FORTRAN-66 and FORTRAN-77 are available in implementations for all mainframe machines as well as for mini- and micro-computers.

The revolution that took place during the last three decades in the field of computational physics and engineering is in equal measure attributable to advances in computer hardware and to the efforts of computer scientists and numerical analysts. The hardware aspect of this revolution is characterized by the transition from the use of the electronic tube in computer circuitry to the application of the transistor, and currently to the use of large-scale integration in the memory chips of computers. The development of code compilers kept in step with the advances in computer hardware.

Life Cycle Costs of Computer Codes

The financial and intellectual investment involved in the development of computer codes used in nuclear engineering is becoming a subject of managerial studies. For computer codes used for many years, the question of code maintainability[(1)] has become closely associated with the concept of code life cycle costing (LCC). This concept views the investment in software development and maintenance as part of a single and continuous operation. In the nuclear field, the maintenance of software in operational condition involves, as a rule, the need to accommodate changes in mainframe hardware.[(2)] Such reprogramming was required, for instance, in the transition from the CDC-6600 machines to CDC-7600, due to the new division of random access memory into Small Core Memory and Large Core Memory. Similarly, the current introduction of the Cray-1 machine involves some reprogramming, which is needed to take advantage of the increase in operating speed made possible by the vectorization features of this machine. Thus, the original cost of software development often becomes a minor part of the total life cycle cost. Average program development costs of $10 per line of program have been quoted.

Some recent studies indicate that the cost of program maintenance may rise to two-thirds of the life-cycle costs;[(3)] other studies indicate that the maintenance costs may be as high as 70 to 90%.[(4)] When code maintenance is done by people other than the code author, the time spent in learning the intricacies of a code written by somebody else adds further to the costs, underscoring the importance of good programming practices and code documentation. To deal with this problem, a permanent position is often created on the staff of the institution originating the code, to support its development and maintenance. Some dedicated professionals devoted a significant portion of their careers to the development and maintenance of one or more of these major computer codes and data bases. Although major competitive advantages accrue to the organizations in question through access to these powerful instruments of engineering analysis, recognition of the endeavors of in-house numerical analysts and programmers generally has been slow in coming. To gain a proper perspective on this problem, it is worth observing that some nuclear codes have been in use for over 25 years, with successive versions appearing every few years. A good example is the PDQ code, which appeared in its first version in 1956 and is still alive and doing well, now in version 8. The development and maintenance of a complex code such as KENO represents an investment in excess of 30 man-years. For major codes, lifetime costs exceeding one million dollars are not unusual.

Code Configuration Control

In installations where many computer codes are developed and used, as a rule a vault is provided for storage, in an appropriate environment, of program source decks and object decks, as well as reports describing the equations programmed in the code and their method of solution. Hard copies of codes in the form of printouts, microfilms or microfiches may be used for storage. It is with reference to these collections that the term "code configuration control" has evolved. Code configuration control is a procedure whereby the main versions of the computer codes used in the design of licensed reactors are documented, verified and stored at the originating institution. The custody of the software record collection is, as a rule, entrusted to a qualified senior programmer. He or she serves as a liaison officer for contacts with the NRC, and is cognizant of all the additions, modifications and extensions of the codes under configuration control.

In the case of codes dealing with the key aspects of reactor safety, such as the emergency core cooling systems (ECCS), the Nuclear Regulatory Commission also requires, for archival purposes, the storage of magnetic tapes with the source decks for the relevant codes. In addition, NRC insists that computer codes used in reactor design and safety work be provided with a verification procedure. To verify the correct operation of the code in question, a person other than the code author or the engineer responsible for code maintenance, carries out calculations and documents the method used. The installation supporting the code is subject twice a year to audits of code verification status by the NRC, with the codes to be audited chosen at random from the local collection.

One effect of the requirement for code configuration control is to freeze the form of codes used for routine design. The rationale for establishing this fairly elaborate system of administrative control is based on the hope of eliminating, or at least minimizing, the ambiguity of results that might arise in the case of repeat calculations or back-checking with different versions of the same code. Without minimizing the advantages of this procedure, it must be recognized that complete freezing of design codes would create a severe disincentive for further progress in code development, and should be used sparingly. Throughout the industry and at the national laboratories there is a continuing effort to write new computer codes in those areas of reactor design or operation that deserve particular attention. Code development efforts are now heavily concentrated on the analysis of all aspects of reactor safety. Attendant on code development is the benchmarking activity described in the following section.

Benchmarking and Quality Control

An aspect of standards and regulations gaining in importance as the industry matures is quality control, in applications involving not only reactor hardware and instrumentation but nuclear engineering software as well. The process of establishing and testing the quality of computer codes and methods depends, in turn, on the selection and application of computational benchmarks.

Progress in code development and attempts at code standardization was, in the past, paralleled by progress in benchmarking; both are gaining in importance of late. These activities grew in response to requirements for quality control and configuration control advocated by the NRC. The Computational Benchmark Problem Committee (CBPC) was formed in the 1960s to develop a collection of analytical problems with mathematically well-defined solutions, against which numerical solutions obtained with various computer codes could be tested. This Committee is currently operating under the chairmanship of B. D. Ganapol, and has produced for core physics problems "The Benchmark Problem Book," originally issued in 1968, and supplemented in 1972 and 1977.[(5)] A new supplement was in preparation in 1983.[(6)] The similar "Shielding Benchmark Problems" was produced at Oak Ridge under the editorship of A. E. Profio.[(7)]

The benchmarks take the form of source situations in well-defined geometries and compositions. For its inclusion in the benckmark problem book, agreement between two independent solutions to the given problem is required. The "Benchmark Problem Book" and its first two supplements contain 15 different source situations, as shown in Table 1. The third supplement, to be issued in 1984, will contain three more source situations. The source situations are intended for testing and intercomparing computer codes. The formulation of the test problems is, as a rule, undertaken by individual investigators, and the problems and their test solutions are then submitted for approval to the CBPC.

From time to time, additional benchmark problems are proposed, and their results published by analysts active in the nuclear field. An example of a benchmark calculation in a multidimensional reactor kinetic analysis, using the alternating direction explicit method of solution with exponential transformation, was presented in 1976 by McDonnell[8] for a Canadian D_2O moderated reactor. Another example of a benchmark calculation, this time for a fast breeder reactor, was given by Buckel, Kufner and Stehle[9] in 1978. In the same year, Lathrop[10] presented a multigroup benchmark problem for the numerical solution of the Boltzmann integral equation. Recently Lee and associates[11] presented a discussion of a two-region diffusion theory benchmark problem in few groups, involving

Table 1. Source Situations in the "Benchmark Problem Book"

1. Small Spherical Critical Experiment (Lady Godiva)
2. High Temperature Gas-Cooled Reactor Configuration
3. A Homogeneous, Two-dimensional Slab
4. A Simple Highly Nonseparable Reactor

---

5. Two-dimensional Isolated Source in an Absorbing Medium
6. Infinite Slab Reactor Model
7. Monoenergetic Point Reactor Model

---

8. Two-dimensional (r-z) Reactor Model
9. Multidimensional (Hex-z) HTGR Problem
10. PWR Thermal-Hydraulics, Cross-channel Flow, Different Heat Flux
11. Multidimensional (x-y-z) LWR Model
12. Neutron Transport in a Cylindrical Black Rod
13. Neutron Transport in a BWR Rod Bundle
14. Multidimensional (x-y-z) BWR Model
15. Neutronic Depletion Benchmark Problems

---

16. Fast Reactor Statics (to be added in 1984)
17. Fast Reactor Kinetics (to be added in 1984)
18. CANDU Reactor Kinetics (to be added in 1984)

space kinetics of water-moderated systems, and of liquid metal fast breeder systems.

Additional benchmark problems for transport theory solutions were presented recently by Westfall[12] and Ganapol.[13] In the latter work, analytical solutions were evaluated to a high degree of accuracy by the multiple collision method, also known as the Neumann series solution, or method of successive approximations. They are applicable to relatively simple problems in transport theory of monoenergetic neutrons or charged particles and other problems. In spite of its versatility and availability of solutions, the method is limited to infinite media and to short times after the initiation of the transport event; at longer times it is numerically ineffective.

Benchmarking of computer codes for stress analysis is carried out under the auspices of ASME, in particular, the Committee for Computing in Applied Mechanics of the Applied Mechanics Division, and the Computer Technology Committee of the Pressure Vessels and Piping Division. Test problems have been formulated,[14] comparisons of computer code results with experimental data have been made,[15] and results obtained with different stress analysis codes have been intercompared.[16] Four international conferences have been held on Structural Mechanics in Reactor Technology (SMiRT) to share results.

Finally, it is worth observing that even though the benchmarking effort is in essence an attempt at standardization, so far computer codes used in the design of nuclear reactors have not been standardized to any significant extent. Nor have the computer codes themselves been used as part of any standard, mostly in recognition of the fact that for the majority of codes there is still room for improvement in the accuracy and speed of computation.

REFERENCES

1. I. M. Soi and K. K. Aggarwal, "A Life-Cycle Cost Viewpoint of Software Maintainability," *Comput. El. Eng.*, 8, No. 4, 277–282 (1981).
2. D. L. Parnas, "Designing Software for Ease of Extension and Contraction," *IEEE Trans. Software Engin.*, SE-5, 128–137 (1979).
3. W. W. Agresti, "Measuring Program Maintainability," *J. Syst. Management,* 33, No. 3, 26–29 (1982).
4. "DOD Wants a Twenty-five-Year Lifespan for its New Languages," *Infoworld,* 4, No. 41, 27 (Oct. 1982).
5. Argonne Code Center, "Benchmark Problem Book," ANL-7416, Argonne National Lab. (1968); Suppl. 1 (1972); Suppl. 2 (1977).
6. R. M. Westfall, "Computational Benchmark Problem Committee—Present Activities and New Problem Areas," Topical Meeting: Advances in Reactor Computations, Salt Lake City (Mar. 28–31, 1983).

7. A. E. Profio, ed., "Shielding Benchmark Problems," ORNL-RSIC-25, (1968); Suppl. 1 (1970); Suppl. 2 (1973).
8. F. N. McDonnell, "D_2O Reactor Kinetics Benchmark Problem," AECL-5377, Chalk River Nuclear Lab. (1976).
9. G. Buckel, K. Kufner, and B. Stehle, "Benchmark Calculations for a Sodium-Cooled Breeder Reactor by Two- and Three-Dimensional Diffusion Methods-II," *Nucl. Sci. Eng.*, 67, 257 (1978).
10. D. K. Lathrop, "Numerical Solution of the Boltzmann Integral Equation—A Multigroup Benchmark Calculation," *Nucl. Sc. Eng.*, 33, 255–257 (1968).
11. Clarence E. Lee, Wesley C. P. Fan, and J. S. Rottler, "Diffusion and Kinetics Analytic Benchmark Calculations," *Trans. ANS*, 44, 280-281, (1983).
12. R. M. Westfall, "Benchmark Solutions for the Infinite Cylinder," *Trans. ANS*, 44, 281–283 (1983).
13. B. D. Ganapol, "Analytical Benchmarks in Time-Dependent Transport Theory via the Method of Multiple Collisions," *Trans. ANS*, 44, 283–285, (1983).
14. J. A. Clinard et al., "Comparison of Typical Inelastic Analysis Predications with Benchmark Problem Experimental Results," pp. 79–98: "Pressure Vessel and Piping—Verification and Qualification of Inelastic Analysis Computer Programs," ASME (1975).
15. J. M. Corum, "Evaluation of Inelastic Analysis Methods," *Trans*. 4th Int. Conf. Structural Mechanics In Reactor Technology, Vol. L, paper 4/1, Com. of the European Communities, Brussels (1977).
16. J. R. O'Leary and Y. A. Patel, "Transient Analysis of Class A Nuclear Piping System: A Comparative Study of Three Large Computer Codes," *Trans*. ASME Conf., Miami Beach, pp. 51–58 (June 1974).

5.2 DEVELOPMENT OF NUCLEAR DESIGN METHODS

Numerical Methods and Reactor Theory

The past three decades have witnessed a revolution in the methods of engineering analysis. The computational aspect of this revolution has overcome the limitations of the formal approach to analysis with its derivation of closed form solutions, as a rule followed by series expansions in onc or more of the existing analytical functions. In the last two decades formal analysis has been gradually replaced by adaptation of the iterative approach of successive approximations to the solution for many categories of problems, including those for which the derivation in closed form had not previously been thought possible. In this approach, the iterative approximations can usually be made to any required degree of accuracy, provided sufficient time and computational effort are spent in arriving at the desired answer. An alternative approach, also introduced by the computational revolution, is the computational experiment, as embodied in the Monte Carlo method. The actual, discrete events taking place in the physical world are mocked up mathematically in the Monte Carlo model, and are followed through, one by one, to their logical conclusion. The averages of the totality of events are then derived, and taken to represent the observable macroscopic response of the physical world.

The transition, from analytical approaches involving rigorous mathematical treatment of idealized material configurations to engineering problems of actual reactor lattices, has engaged the talents of many scientists and engineers during the past three decades, and has produced a substantial crop of monographs, textbooks and handbooks of reactor theory. The earliest and best known of these was the pioneering textbook by S. Glasstone and M. Edlund.(1) Their treatment of slowing down theory based on the so-called Fermi Age formulation initially dominated the field of reactor theory, but was gradually replaced by multigroup methods, which are now used to compute neutron and photon energy spectra.

Derivation of equations describing the energy distribution of particles in interaction with media supporting them was thoroughly covered in two monographs, written respectively by E. A. Guggenheim(2) and L. C. Woods.(3) A major treatise on reactor theory, *The Physical Theory of Neutron Chain Reactors,* was authored in 1958 by Alvin M. Weinberg and Eugene P. Wigner.(4) It was followed by an in-depth study of neutron slowing down and thermalization written by M. M. R. Williams.(5) These two works define and discuss the first and the second fundamental theorems of reactor theory. It is worth recalling here that the first theorem states that the neutron distribution in a reactor is the fundamental solution of the wave equation, positive throughout its volume and vanishing at the extrapolated boundary. The second theorem states that the neutron nonescape probability during moderation in a uniform bare reactor is the Fourier transform of the slowing down kernel. The rest of reactor theory depends for its validity on the proof of these two basic theorems.

The computational aspects of reactor theory were covered in the monograph by R. V. Meghreblian and D. L. Holmes(6) dealing with the derivation of reactor physics equations, while the formulation of these equations for machine calculations was described in a work by Melville Clark and Kent F. Hansen.(7) This last work is particularly relevant to the present guide because it covers the major concerns of the methods on numerical reactor analysis in relation to neutron diffusion, transport and Monte Carlo methods. Three important texts in reactor theory, heavily oriented toward numerical methods of machine computation, appeared in the 1970s. The first of these texts was written by George I. Bell and Samuel Glasstone,(8) the second by Alan Henry,(9) and the last by J. J. Duderstadt and L. J. Hamilton.(10) They supplement and extend the treatment of the subject in earlier works.(11)–(14)

The development of reactor theory has taken place gradually over the last three decades and produced a rich crop of computer codes intended to implement the theory with machine calculations. At each point, the size and complexity of the best of these codes have matched the full capability of digital computers coming on the market. This race between computer hardware and reactor analysis software continues unabated and presents a continuous challenge to all engaged in the promotion and maintenance of the nuclear industry. Later sections of this chapter and the following two chapters briefly describe the most important features of the currently used leading computer codes implementing numerical methods of nuclear and thermo-hydraulic design.

General Description of Numerical Methods

From its very inception, nuclear engineering has grown in close association with the rise and expansion of numerical analysis methods. To assist in the generation of new reactor designs, a large body of computer codes was developed, which now constitutes a valuable national resource. The resulting software of neutronics and photonics incorporates much of the know-how developed in the last quarter century for translating the methods of physical mathematics into the language of computer science. The effort spent in developing this large body of computer codes, when evaluated in terms of man-years invested, constitutes a measurable part of the design cost of the early reactors.

In broad terms, the numerical methods of reactor analysis deal with the prediction of distribution in space, energy, direction, and possibly time, of the reaction rates of neutrons and photons with matter; and with changes in time of the density of the reactants and the products of the resulting transmutation reactions, be they material particles or deposition of energy. This includes the depletion and breeding of fissile materials, the transport and attenuation of radiation through shields, and the generation and removal of heat, with the associated transport and changes of phase, in reactor coolants. The development of advanced methods of numerical analysis for the solution of problems in neutronics and photonics has led to the creation of very powerful computational tools in the form of multidimensional, multigroup computer codes capable of treating the static and the transient effects in reactor theory. In the latter group, transients of short duration associated with power level changes, as well as long transients at constant power level associated with fuel burnup, can now be effectively handled.

Approaches to the numerical treatment of reactor theory fall into two broad categories: the deterministic and the stochastic methods. Under the heading of deterministic methods one usually places the various solutions of the neutron transport equation in either the diffusion or the discrete ordinate form. These solutions are based on aggregate treatment of events of such multiplicity as to be considered individually indistinguishable, and thus amenable to treatment as a continuum. They include various direct calculation approaches to the solution of problems in reactor neutronics and photonics, such as the integral, discrete ordinate, diffusion and nodal methods of solving the neutron or the photon transport equation. Stochastic methods, on the other hand, apply several variants of probabilistic approach to the numerical simulation of physical processes at the individual event level. The aim of this approach is to simulate life-histories of individual particles participating in interactions with atoms of the transport medium, supported by various advanced techniques of life-splitting, and followed by suitable tallying.

Two aspects of numerical computation are of major importance to the analyst: the capability of the method of solution to treat the essential aspects of the physical problem to the required degree of accuracy, and the capability of the computing machinery to handle the size of the problem without compromising its complexity. The first of the above aspects is related to the much broader problem of the existence of finite solutions to the type of equations used to formulate reactor physics phenomena.

Even in the early years of the development of nuclear engineering, it had become apparent to the leading workers in the field of reactor theory that closed form solutions to reactor problems of interest could be expected for only the most simplified geometries and compositions of chain reacting assemblies. Early work in this area was done by R. E. Marshak[(15)] for large plane slab geometry, followed by K. M. Case, F. De Hoffmann and G. Placzek.[(16)] In their classic work, "Introduction to the Theory of Neutron Diffusion," the latter authors derived solutions in closed algebraic form for the simple half-space, infinite slab, infinite cylinder and spherical geometries. Although it was initially intended as a two-volume work, only the first volume was ever issued. The work was continued by B. Davison who, in *Neutron Transport Theory,*[(17)] derived series solutions for similar geometries. His closed form numerical solutions to the required differential and integral equations were obtained from expansion in terms of orthogonal functions appropriate for the problem in hand.

Once the expressions for the closed form solution to each of the simple problems were coded for the computer, a substitution of appropriate values of dimensions and physical properties of the medium into the closed form equations was sufficient to compute the required numerical answer. Apart from the accuracy of the physical data for input, the accuracy of the solution was dependent on the number of terms included in the series expansions derived for the solution, and the physical approximations assumed prior to deriving the solution. The success of this approach was predicated on the ability to derive through mathematical transformations the closed form solutions for the geometry of the problem. Unfortunately, only the simplest geometries could be tackled in this manner; hence the search for other, more flexible approaches to problems dealing with more complex geometries.

Limitations of Machine Capacity

When the solution of physically large problems is to be undertaken, the mesh requirements become formidable and test the capacity of even the largest computers to the very limit. Fortunately, in most cases, advantage can be taken of the symmetries inherent in the geometrical layout of fuel assemblies and the cores of reactors. The computer codes used to analyze them as a rule provide rotational and reflectional symmetries. In the horizontal plane, the reflective (zero current) or periodic (the same current) boundary conditions can be applied along the 1/2, 1/4 or 1/8 core axes of symmetry. In the vertical direction, reflections about the reactor midplane are often possible.

Some computational conditions exist, however, in which advantage cannot be taken of full symmetry. This is the case when the distribution of power in a core with a stuck control rod is to be computed. When the control rod, or rod cluster, is located on the axis of symmetry, 1/2 symmetry boundary conditions can be used, but in the off-axis placement of rod, the full core must be represented. Otherwise, when fuel loading patterns allow it, 1/4 or 1/8 symmetries can be used. The resulting reduction of the computing effort is evident. As an example, in the case of a core containing burnable poison rods replacing some fuel rods, it may be desirable to represent each fuel rod in the reactor explicitly as a separate cell. This translates into 50,625 mesh cells in a large PWR core containing 193 fuel assemblies, each with 225 cell positions. Representation of baffles, water gaps, etc., adds to this requirement.

A specific example of the space required in the random access memory of a digital computer is the storage needed to analyze radiation transport in the Demonstration Tokamak Hybrid Reactor (DTHR).[18] This is a physically large fusion reactor model arranged in a toroidal geometry with a large cavity in its center. The radiation transport problems of DTHR can be analyzed to a different degree of approximation, in geometries shown diagrammatically in Figure 6, by the one-, two-, and three-dimensional transport theory methods described later in this chapter. The mesh requirements of these solutions are summarized in Table 2 where, for the

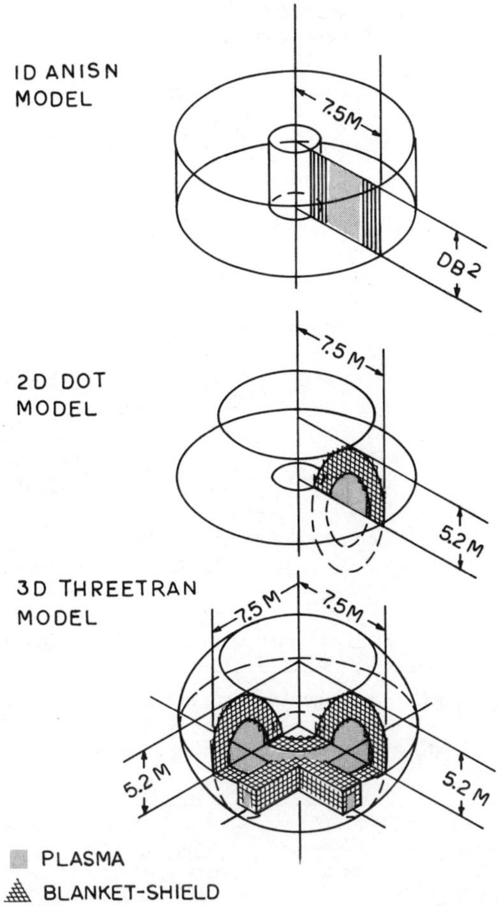

Figure 6. Schematics of code mesh requirements

purpose of radiation transport, the dimensions of the system to be analyzed are expressed in units of transport mean free path of the radiation in question. Also listed at the end of the table are the Small Core Memory (SCM) and the Large Core Memory (LCM) capacity of the CDC-7600 computer in both the decimal and the octal number systems. The table shows that the three-dimensional treatment of the DTHR problem with the THREETRAN code exceeds the present capacity of the CDC-7600.

Table 2 compares the space mesh requirements of three well-known discrete ordinate computer codes: ANISN, DOT and THREETRAN. A discussion of the theory embedded in these codes is given in detail in the work by H. Greenspan, C. N. Kelber and D. Okrent.[19]

Table 2. Computer Memory Requirements for a Large Transport Problem

Model	Dimension of Transport Problem 1–D (ANISN)	 2–D (DOT-III)	 3–D (THREETRAN)
Reactor Dimensions (m)			
D—Depth (radius)	7.5	7.5	7.5
H—Half height	—	5.2	5.2
W—Half width	—	—	7.5
	Number of Space Mesh Points		
Space Mesh Width (cm)	D	D×H	D×H×W
10.0 cm width	75	3,900	292,500
5.0 cm width	150	15,600	2,340,000
2.5 cm width	300	62,400	18,720,000
	Computer Memory Size Required (words)		
Approximation:	$S_4P_1G_{15}$	$S_4P_1G_{15}$	$S_4P_0G_4$
Small Core Memory	31,106	26,108	40,000
Large Core Memory	—	41,033	400,000
	Computer Core Storage Available in CDC–7600 (words)		
Small Core Memory $(53{,}248)_{10} = (150{,}000)_8$			
Large Core Memory $(86{,}016)_{10} = (250{,}000)_8$			

Notes:
S_n = Order of quadrature.
P_l = Order of angular expansion of scattering distributions.
G_i = Order of energy group structure.

The above discussion applies primarily to the popular discrete ordinate method of solving the transport problem. Bengt Carlson defines the transport problem[20] with reference to an arrangement of materials constituting a transport medium, in which two major processes take place in parallel: the undisturbed flow, or streaming, of particles and the interaction of the transported particles with the nuclei of the medium. In this medium the interactive collisions affect the direction, velocity and number of the participating particles.

The main assumption in transport theory is that the collisions do not significantly alter or disturb the materials constituting the medium. When material changes are to be accounted for, an iteration is used between the calculation of particle transport through a medium with constant material properties and a calculation of material density changes, holding the neutron flow constant. These iterations are carried out within a single computer program by its subroutines.

It is worth observing that, in applications designed for shielding analysis, both the deterministic and the stochastic methods of computation have benefited in recent years from the introduction of coupling of the neutronics and photonics aspects of the same problem. The coupling involves inclusion of the prompt gamma emission source due to neutron capture as a proper source term in the group transfer matrix, augmented by additional groups for photon energy transfer. New codes are now also available for treatment of the production and transport of delayed gamma emission due to activation sources produced by neutron irradiation.

Two additional topics are pertinent to this survey: cross section sensitivity studies and comparison of calculations with experiment. The first of these deals with a new, and still developing, field of analysis comprising the methods used to study the effects of cross-section uncertainties on the overall uncertainty in calculations. The second deals with the integral experiments performed specifically as a part of the benchmarking process to test the analytical methods of nuclear engineering. Both of these topics will be discussed in later sections.

The main features of the mechanization of transport, diffusion and burnup codes will be discussed below. A more detailed description of the theory and the programming logic of these codes will be found in reports issued by the originating institution. This is true for all codes listed except ANISN, for which no such report has as yet been issued.

Features Common to Many Codes

In the development of computer codes for reactor analysis, it has become customary for the authors to make up code names as acronyms describing the code's function, e.g., ANISN: Anisotropic S_n Code. It is also customary to capitalize code names and append letters or numerals, Roman or Arabic, to denote various code versions. In the past it was the practice for the code description to include in one document the underlying theory, the code logic, the listing of the program, the input description and a sample problem. As the codes grew longer, the inclusion of the full listing made the report too bulky. In many cases, the input description and the sample problem have been issued separately from the main report. In the case of proprietary codes, the publication of the code listing is usually suppressed.

Because source particles, be they neutrons, photons or charged particles, have their highest energy at birth, the computer codes dealing with the history of their propagation share a common characteristic, namely the replacement of the energy coordinate system by the lethargy ($=\ln[E_{max}/E]$) coordinate system, whose origin is customarily placed at some high energy, say 10 MeV.

The most common task of the major neutronics codes is to solve the eigenvalue problem inherent in the reactor equation. For this reason, the codes listed here and many others share some common features, intended to facilitate the design of nuclear systems matching specific eigenvalue requirements. Thus, after the fundamental eigenvalue of the assembly has been computed for a given input configuration of neutron multiplying system, the codes provide the means for studying the effects on it of variations in the value of the following parameters:

- Material density of a fixed component of a fixed zone.
- Zone width, or depth, for a fixed composition.
- Concentration of a soluble poison throughout the system.
- Effect of temperature change of the system.

Once the fundamental eigenvalue has been obtained, a search can be made for the value of any of these parameters that will produce an eigenvalue matching the input quantity. The first two of the searches listed above are used by designers to produce specifications for the initial configuration of a chain-reacting system with a desired margin of reactivity. The third search is commonly used to obtain the variation in time of the soluble poison concentration required to keep the system just critical until the reactivity reserve present in the unburned fuel drops to zero. The last search feature is available only in the most advanced codes, such as PDQ-7, and enables the designer to plan for a gradual reactor power coast down at the end of the irradiation cycle, and to take advantage of the reactivity margin available in the cores of most present day reactors, as a consequence of the negative temperature coefficient of reactivity.

Pre- and Post-Processing of Throughput Streams

The use of major codes for nuclear reactor design is commonly preceded by the application of the input processing codes, to generate cross-section libraries in the required format, to generate a space mesh, to obtain proper material densities, etc. In some cases the pre-processing codes constitute a part of an integrated code system, such as ARC, CLYDE, or AMPX-II, which will be discussed below. The use of these major codes is also commonly followed by post-processing of the results with one of the minor ancillary codes, to convert the generated flux distributions into quantities of engineering interest, such as burnup, radiation heating rates, material damage rates and time-dependent activation rates. The computer output may be fed into COMPLOT machines to obtain topological maps of isoflux or isopower contours in the domains analyzed.

One aspect of the post processing of computer output is an attempt to control the volume of the paper printout. Since some parts of printouts are rarely used, most codes have options for selecting the detail of output required in the printout. For instance, the full display of angular fluxes generated by transport theory codes is usually requested only at the problem checkout stage. Even with these options, a major reactor study will produce stacks of printout capable of filling the entire office space of the unwary analyst, creating a fire hazard and testing the structural soundness of the supporting floor. To prevent the snow-under problem, most mainframe manufacturers now offer post-processing equipment to record code printouts photographically on microfiches, without going through the paper stage at all. The format of microfiche output is similar to that used for reducing hardcopy technical reports. It gives the analyst the choice of receiving a few squares of plastic instead of a foot-high stack of printouts. This convenience is not without its own problems: meticulous archiving and indexing of the microfiches is a must. A magnifier-reader in good order is required, plus the ability to locate on the microfiche surface the pages and the numbers of interest, without the ability to mark the printout as a guide. The paper envelope holding the microfiche is the only place where any marks or comments can be made.

The proper formatting of output in the post-processing of computer runs has not yet received adequate attention in standard design procedures. Ideally, the numbers, tables or graphs defining the design and appearing in the final report should be formatted for direct inclusion in that report without transcription of any sort. This would eliminate embarrassing errors in transcription, and reduce the retyping required. The lead in implementing these desiderata is now in the national laboratories, where the process of producing reports containing nuclear cross sections, or other nuclear data, has been almost totally computerized.

REFERENCES

1. Samuel Glasstone and Milton Edlund, *Nuclear Reactor Theory,* D. Van Nostrand, Princeton, NJ (1952).
2. E. A. Guggenheim, *Boltzmann's Distribution Law,* Interscience, New York (1955).
3. L. C. Woods, *Introduction to Neutron Distribution Theory,* John Wiley & Sons, New York (1964).
4. A. M. Weinberg and E. P. Wigner, *The Physical Theory of Neutron Chain Reactors,* The University of Chicago Press, Chicago (1958).

5. M. M. R. Williams, *The Slowing Down and Thermalization of Neutrons,* John Wiley & Sons, New York (1966).
6. R. V. Meghreblian and D. L. Holmes, *Reactor Analysis,* McGraw-Hill, New York (1960).
7. Melville Clark, Jr., and Kent F. Hansen, *Numerical Methods of Reactor Analysis,* Academic Press, New York and London (1964).
8. George I. Bell and Samuel Glasstone, *Nuclear Reactor Theory,* Van Nostrand Reinhold Co., New York (1970).
9. Alan F. Henry, *Nuclear-Reactor Analysis,* The MIT Press, Cambridge (1975).
10. J. J. Dunderstadt and L. J. Hamilton, *Nuclear Reactor Analysis,* John Wiley & Sons, New York (1976).
11. A. Radkowsky, ed., "Naval Reactor Physics Handbook," Vol. I Selected Basic Techniques, Naval Reactors, Division of Reactor Development, U.S. Atomic Energy Commission (1964).
12. J. R. Lamarsh, *Introduction to Nuclear Reactor Theory,* Addison-Wesley Publishing Co., Reading, MA (1966).
13. J. L. Meem, *Two Group Reactor Theory,* Gordon & Breach, Science Publishers, New York and London (1964).
14. D. Jakeman, *Physics of Nuclear Reactors,* American Elsevier Publishing Co., New York (1966).
15. R. E. Marshak, "On the Milne Problem for a Large Plane Slab with Constant Source and Anisotropic Scattering," LADC-244, Los Alamos Lab. (1945).
16. K. M. Case, F. De Hoffmann, and G. Placzek, "Introduction to the Theory of Neutron Diffusion," Vol. 1, Los Alamos National Lab. (1953).
17. B. Davison and J. B. Sykes, *Neutron Transport Theory,* Clarendon Press, Oxford (1957).
18. "Conceptual Design of a Demonstration Tokamak Hybrid Reactor," WFPS-TM-107, Westinghouse Electric Co., Pittsburgh, PA (1978).
19. H. Greenspan, C. N. Kelber, and D. Okrent, eds., *Computing Methods in Reactor Physics,* Gordon & Breach, Science Publishers, New York (1968).
20. Bengt G. Carlson, "Numerical Formulation and Solution of Neutron Transport Problems," LA-2996, Los Alamos Scientific Lab. (1964).

5.3 INTEGRATED COMPUTER CODE SYSTEMS

In addition to individual codes used for computing particular aspects of reactor physics or performance, the various national laboratories, the Nuclear Regulatory Commission and the Electric Power Research Institute as well as major reactor suppliers have all developed highly integrated computer code systems. These systems are distinguished by the compatibility of input and output formats, the incorporation of built-in subsets of the data base and various sequence controllers.

The best-known of these integrated code systems are: the SCALE system,[1] developed for NRC by the Oak Ridge Laboratory for performing standardized analyses for licensing evaluation; the AMPX system, developed by the Oak Ridge Shielding Information Center for preparation of cross-section libraries needed in fusion work; the Argonne Reactor Computation (ARC) system;[2] and the CLYDE system of the Lawrence Livermore Laboratory. The Electric Power Research Institute is developing its own ARMP system.[3] Major reactor vendors have their own proprietary code systems. The code systems in public domain are discussed in more detail below.

Since these code systems grew out of the early neutron and photon cross-section preparation schemes, requiring, as a rule, sequential application of several processing codes to nuclear data supplied by the experimentalists, a brief mention of the early cross-section sets is in order first.

Early Schemes for Generating Cross-Section Sets

In the early history of the nuclear industry, the preparation of neutron and photon cross section sets needed for given design tasks was done ad hoc, as the need arose. Different cross section input data sets were used, and their treatment was nonuniform, depending greatly on the computational resources available to a given analyst. Awareness of differences in results has led to attempts at intercomparison of cross-section sets generated in the United States and in Europe. One such early intercomparison is given in a compilation prepared for the European Nuclear Energy Agency by P. L. Olgaard.[4]

To reduce the degree of variability in nuclear data treatment, often occurring even between analysts at the same laboratory, various laboratories began to publish cross-section sets processed by a standard method to a standard format. The earliest was the 18-group set[5] published in 1959 by C. B. Mills of Los Alamos Laboratory, followed shortly thereafter by the 6-group and 16-group sets[6] generated for fast and intermediate reactors by Gordon E. Hansen and William H. Roach. This latest set of cross sections was checked out by transport theory calculations of just critical chain reacting assemblies in simple geometries. The set was included among several others in report ANL-5800,[7] "Reactor Physics Constants." Because of its relatively thorough documentation, the Hansen–Roach set has enjoyed unflagging popularity with the nuclear community to the present day, in spite of other, more extensive cross section libraries appearing later.

The approach taken by Hansen and Roach in preparing their cross-section set was to precompute resonance self-shielded cross sections for all important members of the Th, U and Pu isotope chains, to cover in a sufficient number of steps the range from infinitely dilute to the highly self-shielded condition. The shielding parameter used was the magnitude of the scattering cross section per absorber atom. All cross sections were evaluated at the same temperature and tabulated.

The Hansen–Roach cross section set has retained its popularity with nuclear analysts for the past 20 years, and is still in use as part of larger nuclear computation schemes, where it often forms a system-resident cross-section library, as, for instance, in the KENO Monte Carlo program, or in the space reactor computing scheme of Lawrence Livermore Laboratory. It has also led the Russians to add one more tabular variable, the temperature, and extend the group structure to 26 groups, while applying the concept to calculations of group constants for fast reactors. The report describing the Russian scheme became available in this country through an Israeli translation of a Russian report authored by Abagyan, Bazazyants, Nikolaev and Bondarenko,[(8)] with Bondarenko giving his name to the scheme of cross-section tables used with double interpolation in the degree of self-shielding and temperature.

The computerized version of the Bondarenko double interpolation scheme is more efficient in the use of computer time than repeated evaluation of cross section sets from scratch. Of course, the initial computational effort needed to develop the temperature and self-shielding scale factors is unavoidable, and is quite considerable. In its mature form, therefore, the Bondarenko scheme consists of a set of cross sections for each isotope of interest, unshielded and uncorrected for temperature effects, plus a table of scale factors for each isotope with a resonance cross section, tabulated by self-shielding and temperature. The degree of self-shielding is again measured by the magnitude of the scattering cross section per absorber atom.

The Bondarenko scheme was incorporated into the 1-DX code for cross-section handling developed at Hanford Laboratories for fast breeder work, and is also part of the AMPX scheme developed at Oak Ridge Laboratory for the generation of cross sections for fusion work. Both codes are now a part of much more elaborate integrated computational schemes for handling nuclear analysis. These and other computational schemes will be described in more detail below.

Another popular 16-group cross-section set, for work with fast reactors, was developed at the Argonne National Laboratory by Yiftah, Okrent and Moldauer.[(9)]

The Oak Ridge AMPX-II System

The AMPX-II System[(10)] is a modular code system for the generation and manipulation of coupled neutron and gamma-ray cross sections. The system was developed at the Oak Ridge National Laboratory with the cooperation of other national and industrial laboratories involved in the controlled thermonuclear reactor development program. The intent was to provide the fusion community with a standardized method of generating data, primarily for the neutronic and photonic design of fusion devices, but useful for fission reactor applications as well.

The system consists of linked modules that perform in series the operations needed to transform nuclear data contained in the Evaluated Nuclear Data File (ENDF) to the format required by the various transport theory codes. In the process, corrections are applied to the basic data for temperature and self-shielding effects, using the Bondarenko scheme of correction factors.

The modules of the system, and their functions, are listed in Table 3.

The system has undergone a period of testing at laboratories contributing to its development, is now in operation, but further improvements are still being made to it. An additional post-processor code CTART processes Monte Carlo output.

The Lawrence Livermore CLYDE System

The CLYDE System[11] was developed at the Lawrence Livermore Laboratory by R. J. Doyas, R. E. Dye, R. J. Howerton and S. T. Perkins to handle a whole range of cross-section processing needs. The code takes the ENDF/B library data and transforms them to the ENDL format used in codes developed at Lawrence Livermore Laboratory. It processes neutron interactions and gamma-ray production from neutron-induced reactions.

The data are processed for specified energy group structure, weighting fluxes and isotopes. The code gives flux-averaged cross sections, transfer probabilities, particle and isotope production cross sections, and isotope destruction cross sections plus such other quantities as the number of neutrons per fission. The output data come in a format that can be used directly in discrete ordinate transport codes, diffusion theory codes or stochastic theory codes for computing neutronic and photonic aspects of reactor design, or for carrying out fuel burnup calculations.

Although both the input data and the evaluated library are arranged by increasing energy, the code inverts all output arrays so that numbering and listing of all groups and cross sections conform to the conventional arrangement of energy groups in neutronic and photonic codes; i.e., numbering starts with the highest energy group and continues in order of increasing lethargy.

The system preprocessor codes are listed in Table 4.

The limitations of the code are as follows:

- The flux spectrum must span the group bounds for all orders of angular expansion of cross sections,
- For endoergic reactions, the energy threshold should coincide with the initial energy of the associated cross section list. If it does not, the code will set the initial energy to that value; it will also set the cross section at that energy to zero.

Table 3. Modules of the AMPX System

Name	Function of Module
DRIVER	is the AMPX module manager
NPTXS	prepares point cross sections for material with resonance parameters
XLACS-2	produces multigroup cross sections from ENDF-formatted data (or NPTXS processed data)
LAPHNGAS	produces multigroup secondary gamma-ray production data (photon multiplicities or cross sections)
SMUG	produces multigroup gamma-ray interaction cross sections
CHOX	performs cross section interface management, e.g., combines output from XLACS-2, LAPHNGAS and SMUG
NITAWL	performs resonance self-shielding and produces working libraries
ROLAIDS	treats resonance self-shielding in multiregion geometry
BONAMI	performs resonance self-shielding using Bondarenko factors
CHOXM	generates a self-shielded AMPX master library using SPHINX generated self-shielding factors
XSDRNPM	provides one-dimensional S_n capability for spatial cross section weighting
AIM	provides Binary-Coded-Decimal to Binary conversion of AMPX master libraries
AJAX	merges, collects, assembles, records, joins and/or copies selected data from AMPX master libraries produced by XLACS-2, LAPHNGAS and SMUG
DIAL	edits data from AMPX libraries
VASELINE	plots multigroup and/or point data for visual comparisons
RADE	checks AMPX master libraries for consistency and reasonableness
MALOCS-2	collapses AMPX master libraries with input spectra
PAL	punches reaction cross sections from AMPX master libraries
DOSE	calculates multigroup neutron and gamma-ray dose factors
UNITAB	selects portions of AMPX master libraries to prepare hybrid master sets
CLAROL	replaces cross sections on an AMPX master library
COMET	corrects selected portions of AMPX master libraries
LAVA	converts an ANISN-formatted library into an AMPX working library
CONVERT	converts old XSDRN-formatted library into an AMPX master library
REVERT	reverts AMPX master library to XSDRN format
CONTAC	converts AMPX working libraries into ANISN or CCCC ISOTXS format
GERITOL	places group constants into a "string" library

Table 4. Modules of the CLYDE System

Name	Function of Module
UPDATE	converts angular fluxes to Legendre coefficients, incorporating fitting and thinning procedures
SOPHIST-III	calculates Doppler broadening, including boundary motion and self-shielding of resonance reactions
BONNIE-1	processes scattering kernels for the more time-consuming multibody transfer processes [e.g. (n,2n)]; generates the correlated energy-angle Legendre expansions of these kernels
FLANGELL	generates thermal scattering kernels from S(α, β) data
UPDATE	generates differential cross section data from Legendre expansion coefficients
FISPEC	generates energy tables for the fission transfer kernel
ENDEP	calculates average energy depositions from neutron-induced reactions
ENDFLLL	generates "cold" cross section data from resonance parameters (resolved and unresolved)

The NRC SCALE System

SCALE[12] (Standardized Computer Analyses for Licensing Evaluation) is a multifaceted computational system, designed to provide a standard analytic tool to evaluate for licensing purposes the nuclear design of fuel facilities and shipping packages for radioactive materials. The system consists of several automated analytical sequences, carried out by appropriate code modules, capable of evaluating nuclear criticality, shielding requirements and heat transfer requirements of equipment to be licensed. A minimum of user-specified input is required. The functional modules executing the computational sequences of the system can also be used in a stand-alone fashion.

The components of SCALE provide the following capabilities:

- System driver minimizing the requirements of computer time and magnetic core size.
- Control modules operating with a minimum of input.
- Coupling of data processing and execution of analytical sequences.
- Self-contained and thorough documentation.
- Use of well proven modules and data libraries.

The user selects the analytical sequence required for criticality, shielding or heat transfer and specifies the geometry of the calculation. Input is in a free-form format, and consists of standard engineering parameters. The execution of the analytic sequences is, as a rule, preceded by data preparation, either through cross-section processing or through access to the material composition and property libraries. Control modules of the system are listed in Table 5, functional modules in Table 6, and support modules in Table 7.

The performance and accuracy of the SCALE predictions of criticality of various systems were tested against the results of 25 recently performed critical experiments,[13] using three built-in cross-section sets: a

Table 5. Control Modules of SCALE

Name	Function of Module
CSAS1	One-dimensional criticality safety analysis
CSAS2	Multidimensional criticality safety analysis
CSAS3	Optimum concentration criticality safety analysis (in preparation)
CSAS4	Optimum pitch criticality safety analysis
SAS1	One-dimensional shielding analysis (in preparation)
SAS2	Coupled One-dimensional depletion and shielding analysis
SAS3	Monte Carlo shielding analysis
SAS4	Coupled depletion and Monte Carlo shielding analysis (in preparation)
SAS5	Fuel cask Monte Carlo shielding with automated biasing (in preparation)
HTAS1	Two-dimensional heat transfer analysis of fuel casks

Table 6. Functional Modules of SCALE

Name	Function of Module
BONAMI-S	Resonance self-shielding by the Bondarenko method
NITAWL-S	Resonance self-shielding by the Nordheim method
XSDRNPM-S	One-dimensional discrete ordinates transport analysis
XSDOSE	Calculation of fluxes and dose rates outside a shield
KENO-IV/S	Monte Carlo criticality analysis
COUPLE	Processing of cross sections and spectra for ORIGEN-S
ORIGEN-S	Calculation of isotopic accretion and depletion in fuel
ICE-S	Generation of macroscopic transport group constants
MORSE-SGC/S	Neutron and photon transport by Monte Carlo method
HEATING6	Multidimensional finite-difference heat conduction
KENO-V	Criticality by Monte Carlo method
JUNEBUG	Three-dimensional geometry plotting

Table 7. Support and Property Modules of SCALE

Function of Module
System driver
System subroutine library
Free-form reading routines
Neutron cross-section libraries
Neutron–photon cross-section libraries
ORIGEN-S data libraries
Materials properties libraries
Standard composition library
MARS: combinatorial geometry for multiple array systems
FIDO input system
Interactive input processor

27-group ENDF/B-IV, a 123 group GAM-THERMOS and a 16 group Hansen–Roach set. Satisfactory results were obtained with all three sets. As could be expected, the 123 group set gave the best agreement with experiments.

In a typical safety calculation, using the Criticality Safety Analytical Sequence No. 2 (CSAS2), the driver module, control modules, functional modules and data base all come into play. A single unified set of input data is read by the control module: auxiliary calculations, such as determination of the Dancoff correction, are carried out, cross sections are processed for the given problem with the BONAMI and NITAWL functional modules: and finally the system multiplication factor is computed with the KENO code.[(14)–(16)]

The Argonne ARC System

ARC, the Argonne Reactor Computation System,[(17)] was developed by the Argonne National Laboratory between 1967 and 1978 as the Laboratory's own integrated system for reactor computations. It consists of several related modules,[(18)–(21)] all using a similar format for input, permitting linked calculations. Fast reactors can be analyzed with the system, and hexagonal fuel element geometries can be accommodated.

The system consists of the following components:

- ARC-SYSTEM: a collection of modules that initialize the operation of the system, process RBCDIC input data into the ARC System data sets for use in other computational modules, and provide communication linkage between ARC System interface data sets.

- ARC INPUT Processor: component that reads and analyzes the card image data defining reactor geometry, mesh structure, region and composition assignments and boundary conditions. Specifications for criticality searches and external source conditions are also processed. Insufficient or inconsistent data are identified and tagged. Edits of data sets are produced on request.
- ARC-XSEC1: module that manipulates multigroup microscopic cross-section data to modify isotopic cross-section components, collapse the energy group structure, and homogenize mixtures for use by the neutronics modules of the system.
- ARC-DIF1D: four modules that give one dimensional solutions of the multigroup neutron diffusion equations, execute searches, provide the required edits.
- ARC-DARC2D: module that performs two dimensional solution of the multigroup neutron diffusion equation, executes searches, provides required edits.
- ARC-REBUS2: system of codes for the fuel cycle analysis of a fast reactor with either an equilibrium condition or startup condition with the explicit cycle-by-cycle fuel management. Adjustment of burn cycle time, fresh fuel enrichment or control poison density can be specified.

The bulk of the ARC System software is programmed in FORTRAN-IV and is operational on an IBM 360/195 machine. At least 400k bytes of random access memory are required by the largest module of the system, REBUS2.

REFERENCES

1. J. A. Bucholz, "SCALE—A Modular Code System for Performing Standardized Computer Analyses for Licensing Evaluation," 3 vols., NUREG/CR-200 (Oct. 1981).
2. B. J. Toppel, ed., "The Argonne Reactor Computation (ARC) System," ANL-7332, Argonne National Lab. (1967).
3. "Advanced Recycle Methodology Program, Project Status Report," EPRI-118-1, Nuclear Associates International (1975).
4. P. L. Olgaard, "Nuclear Constants Used in Two-Group Calculations," EACRP-U-21, European Nuclear Energy Agency, Paris, France (1967).
5. C. B. Mills, "Neutron Cross Sections for Fast and Intermediate Nuclear Reactors," LAMS-2255, Los Alamos Lab. (1959).
6. Gordon E. Hansen and William H. Roach, "Six and Sixteen Group Cross Sections for Fast and Intermediate Critical Assemblies," LAMS-2543, Los Alamos Lab. (1961).
7. L. J. Templin, ed., "Reactor Physics Constants," ANL-5800, 2nd edit., Argonne National Lab. (1963).

8. L. P. Abagyan, N. O. Bazazyants, I. I. Bondarenko and M. N. Nikolaev, "Group Constants for Nuclear Reactor Calculations," Consultants Bureau, New York (1964).
9. S. Yiftah, D. Okrent, and P. Moldauer, *Fast Reactor Cross Sections—A Study Leading to a Sixteen Group Set,* Pergamon Press, Elmsford, NY (1960).
10. "AMPX-II, A Modular Code System for Generating Coupled Multigroup Neutron-Gamma Libraries from ENDF/B, PSR-63 Package," Radiation Shielding Information Center, Oak Ridge National Lab. (1978).
11. R. J. Doyas, R. E. Dye, R. J. Howerton, and S. T. Perkins, "CLYDE, A Code for the Production of Calculation Constants from Nuclear Data," Vol. V-A, UCRL-50400, Lawrence Livermore Lab. (1975).
12. "SCALE—A Modular Code System for Performing Standardized Computer Analyses for Licensing Evaluation," Vols. 1-3, NUREG/CR-0200 (Jan. 1982; rev. June 1983).
13. A. M. Hathout, R. M. Westfall, and H. L. Dodds, "SCALE System Cross Section Validation for Criticality Safety Analysis," *Trans. ANS,* 35, 281–283 (Nov. 1980).
14. C. V. Parks et al., "Standardized Analyses of Nuclear Shipping Containers," *Proc.* 7th Int. Symp. Transp. Radioactive Materials, New Orleans, LA (May 15–20, 1983).
15. R. M. Westfall and L. M. Petrie, "Standardized Safety Analysis of Nuclear Fuel Shipping Containers," pp. 824–849, *Proc.* 5th Int. Symp. Transp. Radioactive Materials, Las Vegas, NV (May 7–12, 1978).
16. R. H. Odegaarden, "Application of SCALE System to Nuclear and Thermal Review of Radioactive Material Package Designs," *Proc.* 5th Int. Symp. Transp. Radioactive Materials, Las Vegas, NV (May 7–12, 1978).
17. L. C. Just, H. Henryson, A. S. Kennedy, S. D. Sparck, B. J. Toppel, and P. M. Walker, "The System Aspects and Interface Data Sets of the Argonne Reactor Computation (ARC) System," ANL-7711, Argonne National Lab. (1971).
18. E. A. Kovalsky, J. Zapatka, H. Henryson, J. Hoover, and P. M. Walker, "The ARC System Cross Section Homogenization and Modification Capabilities," ANL-7714, Argonne National Lab. (1971).
19. D. E. Neal, G. K. Leaf, and A. S. Kennedy, "The Arc System One-Dmensional Diffusion Theory Capability, DARC1D," ANL-7715, Argonne National Lab. (1971).
20. T. A. Daly, G. K. Leaf, and A. S. Kennedy, "The ARC System Two-Dimensional Diffusion Theory Capability, DARC2D," ANL-7716, Argonne National Lab. (May 1972).
21. R. D. Hosteny, "The ARC System Fuel Cycle Analysis Capability, REBUS-2," ANL-7721, Argonne National Lab. (Oct. 1978).

6. REPRESENTATIVE CODES FOR NUCLEAR DESIGN

6.1 SELECTED DISCRETE ORDINATES CODES

The Discrete Ordinates Transport Method

Early numerical methods of computation of radiation transport were based strongly on mathematical formulations derived by astrophysicists to describe physical processes taking place in the interior of stars.[(1)] Closed form formulations used in that branch of physics dealt mainly with large, homogeneous and weakly absorbing media for which radiation transfer at the surface could be treated in the infinite half-plane approximation.

When the limitations of the closed form approach could no longer be ignored, an alternate approach was adopted. More complex geometries were tackled by discretizing the Euclidean space through a finite space mesh, and the continua of energy, velocity and direction were discretized similarly. Basic physical equations were then applied iteratively at each discretized point in the multivariate space, and the solution propagated from point to point by requiring the conservation of quantities computed at each mesh point. Thus the discrete ordinates approach was born. More will be said about it below. Here, let it suffice to say that the solutions required by reactor theory can now be machined to the desired accuracy by application of more machine time to the iterative process.

When applied to situations existing in the core of a nuclear reactor, the type of problem described above could be handled by the diffusion approximation. Early reports on this topic were authored by Bengt Carlson[(2),(3)] and described work done at the Los Alamos Scientific Laboratory in the 1940s and 1950s. The first multigroup S_n transport code was written in 1955 at Los Alamos by J. Bendt. In 1958 Gerald Goertzel, and Bengt G. Carlson with G. I. Bell, published the first descriptions of the method of discrete ordinates,[(4),(5)] to be followed in 1960 by Janusz R. Mika and R. Zelazny.[(6)] C. E. Lee gave a full description[(7)] of the S_n approximation to the transport equation in 1962. Further reports authored first by Carlson, then by Kaye D. Lathrop, and finally one authored jointly by Carlson and Lathrop, laid a solid groundwork for the method.[(8)–(10)]

The application of the discrete ordinates method to more complex geometries and compositions made it possible to treat problems for which there were until then no known closed form solutions. Now achievement of convergence became the main aim, i.e., the ability to insure that there was a value to which the iterative solution would converge. It is in the nature of the iterative solutions of the discrete ordinate problems that their convergence is sensitive to the size of the mesh in energy, space and direction, or time, as the case may be. The finer the mesh, the closer the final result to the expected converged solution. On the other hand, the finer the mesh, the bigger the computer memory size required to treat the problem, and greater the computational effort.

The discrete ordinates transport method is capable of handling the analysis of particle transport in situations where diffusion theory breaks down because of its inherent weaknesses, i.e., in transport of radiation through voids, through interfaces between strongly and weakly absorbing media, etc. The method uses the multigroup approach as a rule, combined with a higher order expansion of the scattering distributions. It requires the preparation of proper multitable cross-section and quadrature sets, and careful attention to the correct choice of solution.

A separate branch of analysis integral to the discrete ordinates methods is the generation of quadrature coefficients and direction cosines.[(11)] These two sets of input data are needed for each transport calculation and are obtained after a decision has been made regarding the order of quadrature to be used, i.e., the number and relation between the discrete ordinate directions chosen for the given analysis. In this subsidiary calculation, the surface of the unit sphere, imagined to be centered on the calculation point of the space mesh of the problem, is divided into segments, each pierced by one direction vector. Here various relations are possible between the chosen direction vectors, to produce half-symmetric, fully symmetric or forward-biased sets, the choice being dictated by the physics of the problem under consideration.

Powerful as it is, the solution of the transport equation in the discrete ordinate S_n approximation nevertheless has its own problems:

- The onset of ray effects.
- The possibility of generating negative fluxes.
- Oscillations in the flux solution attributable to the diamond-difference flux calculation scheme.
- Limitations of modeling curvilinear boundaries by rectangular mesh.

"Ray effects"[12] are often observed in the analysis of radiation transport through media of low transport cross section when conventional symmetric low quadrature techniques are used. These effects are purely numerical, and produced through the constructive and destructive superposition of contributions to the scalar flux from the flux components along each of the discrete directions. Methods were developed[13]–[15] to remedy this effect, for instance, with Lagrangian interpolation in the polar angle, in post processing of results.

The analyst must be on the lookout for the appearance of these problems in situations where they are likely to arise, and prevent their appearance by paying proper attention to the choice of the analytical approach, the size of the space mesh, the selection of the flux solution mode, etc. Of particular importance here is the choice of the size of the space mesh relative to the optical opaqueness of the medium for neutrons and photons. The characteristic "optical" dimension is the transport mean free path for the radiation in the medium, measured in reciprocals of the macroscopic transport cross section.

The magnitude of the transport cross section is energy-dependent and varies from group to group. In strongly absorbing media, in which the contribution from the absorption cross section to the transport cross section dominates, the transport mean free path is shortest in groups of lowest energy. These groups, therefore, set the mesh requirements. In problems involving deep penetration of neutrons, the mesh size requirements based on low energy cross sections can be relaxed somewhat, and the mesh selected relative to high energy groups, since these groups dominate in deep penetration. In two-dimensional problems, the equalization of mesh size in both dimensions will often help to avoid the problems.[16]

The last limitation of the discrete ordinates method concerns the innate inability of the method to give true values of currents at interfaces, when rectangular mesh is used in modeling curvilinear shapes. No matter how fine the mesh, the boundary surface is always an order of the square root of 2 larger than the real surface. Thus, although the number of particles leaving a given zone will be computed correctly, the ratio of particles to area crossed (the current) will be correspondingly faulted. When more accurate values of currents are required, and the problem geometry permits it, the use of triangular, rather than rectangular, mesh is indicated.

The DTF-IV and DTF-BURN Codes of Los Alamos

The DTF-IV Code[17] for calculations of neutron and photon transport was developed at the Los Alamos Scientific Laboratory by K. D. Lathrop on the basis of a theory formulated by Bengt Carlson for the discrete ordinates method. With the DTF code the mature form of discrete ordinates transport codes was established. Its present version is an improvement over earlier versions with respect to speed and accuracy of solution, and is representative of a larger group of codes in this category.

In the DTF code the multigroup, one dimensional Boltzmann transport equation is solved for slab, cylindrical or spherical geometries in the discrete S_n approximation. Isotropic or linearly anisotropic scattering is permitted between all energy groups. The anisotropic scattering is represented by the Legendre polynomial expansion of the differential scattering cross section, which is truncated after a few terms. Complete scattering transfer matrices in energy are allowed for as many expansion coefficients as are permitted by machine storage limitations. An optional, pointwise convergence test is incorporated, as well as a neutron-conserving correction routine to prevent the occurrence of negative fluxes.

The code incorporates an upscatter scaling procedure that accelerates and improves convergence in problems involving scattering from low to high energies, in reverse of the usual situation. This is needed in the case of neutrons returning to a hot reactor core, after an initial moderation from high birth energies in a low temperature reflector. Other features of the code include the capability for:

- Fixed distributed source calculations
- Reactivity calculations
- Time absorption
- Material concentration search
- Zone width search
- Outer radius search

The limitations of the code include lack of the shell source input option, and a rather short output information format.

A burnup version of the code, DTF-BURN, is also available, for point or zone depletion of either uranium or thorium fuel chains.

The DTF-BURN Code[(18)] was developed by J. J. Prabulos of the Los Alamos Scientific Laboratory by coupling the SIZZLE code[(19)] with the DTF-IV transport theory code. In its present version, the code calculates pointwise isotopic depletion using fluxes computed by the DTF method.

Additional features of the code include:

- Eigenvalue calculation
- Burnup period search
- Element concentration search
- Zone thickness search

A different search option can be selected at the beginning of each time step. A nuclide concentration search can be executed directly on atom densities of a specific number of materials, and an option exists for keeping the total number of heavy atoms per region constant. The number of time intervals for burnup, and the length of each interval, are specified by the user. Zones, in which burnup calculations are to be performed, are specified by the user. Microscopic cross sections can be entered by region, so that spectral effects can be accounted for. Uranium or thorium depletion chains can be handled.

In its present form, the restrictions of the DTF-BURN code are:

- There is no fixed source option. The equations for it are built into the code, but the storage space has been eliminated. This feature can be restored with a reasonable amount of effort.
- In the equations built into the code, higher isotope production through the (n,2n), (n,3n), (n,α) and (n,p) reactions is not accounted for. This is a more serious deficiency, important in cases when the code is applied to the analysis of burnup media supporting hard neutron spectra.

This code represents the first attempt to furnish a transport theory code with a capability for handling burnup problems.

The ANISN and TDA Codes of Oak Ridge

The ANISN Code[20] was developed by Ward W. Engle, Jr., of Oak Ridge National Laboratory for the solution of the neutron and gamma-ray transport problems by the deterministic discrete ordinates method. Several versions of the program written in FORTRAN-IV are now in existence, with modifications developed by various laboratories[21] to adapt it for use on various computers, including such machines as the CDC-7600, CRAY-1 and UNIVAC-1108. Even though this is one of the most heavily used transport codes, no formal code description has ever been issued by its author, the user's manual serving so far as its only description.

In the code, the Boltzmann transport equation is solved iteratively in one dimension for slab, cylindrical or spherical geometry in the discrete S_n approximation. Anisotropic scattering is permitted between all energy groups. In addition, a diffusion theory solution may be obtained for any, or all, energy groups.

For input, the code requires cross sections preprocessed in the multitable, multigroup format,[22] and computes the multigroup directed fluxes and currents, as well as the direction-integrated scalar flux. Because of the multiplicity of fluxes computed, the computation effort in the transport theory option is generally an order of magnitude higher than that required by the diffusion theory calculation for a comparable energy group structure.

The types of problems that can be solved with ANISN are as follows:

- Fixed source calculations, with distributed, or shell source
- Reactivity calculation
- Time absorption calculation
- Material concentration search
- Zone width search
- Outer radius search
- Transverse buckling search

The present versions of the code provide for variable dimensioning with regard to the number of:

- Energy groups
- Angular components of cross sections
- Direction components of computed fluxes
- Isotopes and mixes
- Space mesh points and regions

The combinations of values of the above parameters that are chosen for input to a particular problem are limited only by the capacity of the computer used and the availability of suitably formatted cross sections. Expressions for relating the size of the ANISN problem to the size of the random access memory of the computer used for executing the problem exist for all versions of ANISN. The choice of input values usually requires a compromise between the accuracy desired and the budget available for computing.

Coupled neutron–photon transport problems can be solved simultaneously with ANISN using suitably prepared cross-section sets.

A variety of methods is available for computing mesh fluxes:

- Linear diamond difference, no negative flux fix up
- Linear diamond difference, with step function flux fix up
- Step function, no negative flux fix up
- Weighted difference, no negative flux fix up
- Linear diamond difference, with weighted difference flux fix up

Activity production rates can be computed for reactions included in the cross-section library in extra positions at the head of the cross-section table. Group collapsing of fluxes and cross sections for simplification of energy group structure is also available as an option in the code. A number of service codes are available for post-processing of ANISN output, capable of computing heating rates, dose rates, flux plotting, etc.

The code was thoroughly tested in various applications, and the results obtained were verified against results of calculations with other codes, and against experiments.[23] During its use in the NERVA rocket project, computing capabilities of the code were extended as a result of supporting research.[24]

A time-dependent version of the ANISN is also available,[25] known as the TDA Code. Intended for calculation of the movement of a particle wave through a transport medium, the code has been applied to the transport of both gamma rays and neutrons. The code has strict requirements for the relation of the time mesh and the space mesh, namely, that the distance traveled by the incident wave during one time mesh be no greater than the width of one space mesh. In most physical situations this implies time steps on the order of nanoseconds, and computations limited to short-lived phenomena.

TDA allows the choice of two types of source and one initial condition. The first type of source is the space and energy distributed source with a step function time distribution; set equal to zero after the first time interval. Alternatively, an analytic first-collision source can be specified to represent either a delta function, or a time-dependent isotropic source. The source geometries available are: a sphere, an infinite plane, a monodirectional source in a slab geometry, as well as infinite-line source in a cylinder. As an initial condition, the complete flux distribution in space and angle can be specified at zero time, either alone or in combination with one of the two sources described above.

REFERENCES

1. S. Chandrasekhar, *Radiative Transfer,* Oxford University Press, Oxford (1950).
2. Bengt Carlson, "Neutron Diffusion Theory, Integral Theory Methods and Formulas," AECD-2835, Los Alamos Scientific Lab. (1949).
3. Bengt Carlson, "Neutron Diffusion Theory—The Transport Approximation," AECU-725, Los Alamos Scientific Lab. (1950).
4. G. Goertzel, "The Method of Discrete Ordinates," *Nucl. Sci. Eng.*, 4, 581–587 (1958).
5. B. G. Carlson and G. I. Bell, "Solution of the Transport Equation by the S_n Method," *Proc.* of the Second U. N. Conf. on Peaceful Uses of Atomic Energy, Vol. 16, Geneva, Paper P/2386, 535–569 (1958).
6. J. Mika and R. Zelazny, "Polynomial Approximation in Neutron Transport Theory," *Bull. Acad. Polon. Sci., Ser. Sci. Math. Astron. Phys.*, 8, 59 (1960).

7. C. E. Lee, "The Discrete S_n Approximation to Transport Theory," LA-2595, Los Alamos Scientific Lab. (1962).
8. Bengt G. Carlson, "Numerical Formulation and Solution of Neutron Transport Problems," LA-2996, Los Alamos Scientific Lab. (1963).
9. Kaye D. Lathrop, "Anisotropic Scattering Approximations in the Boltzmann Transport Equation," LA-3051, Los Alamos Scientific Lab. (1964).
10. B. G. Carlson and K. D. Lathrop, "Transport Theory—The Method of Discrete Ordinates," LA-3251-MS, Los Alamos Scientific Lab. (1965).
11. J. P. Jenal, P. J. Ericson, W. A. Rhodes, D. B. Simpson, and M. L. Williams, "The Generation of a Computer Library for Discrete Ordinates Quadrature Sets," ORNL-TM-6023, Oak Ridge National Lab. (1975).
12. D. K. Lathrop, "Ray Effects in Discrete Ordinates Equations," *Nucl. Sci. Eng.*, 32, 357–369 (1968).
13. R. K. Disney, J. Jedruch, R. G. Soltesz, R. S. Kaiser, and S. L. Zeigler, "Annual Report for the Contract Year 1970, Supporting Research and Technology, Project 493," pp. 6-1 to 6-28 in WANL-TME-2778, Westinghouse Astronuclear Lab., Pittsburgh, PA (1971).
14. K. D. Lathrop, "Remedies for Ray Effects," *Nucl. Sci. Eng.*, 45, 255–268 (1971).
15. W. F. Miller and W. H. Reed, "Ray-effect Mitigation Methods for Two-dimensional Neutron Transport Theory," *Nucl. Sci. Eng.*, 62, 391–411 (1977).
16. J. Jedruch, "Transport Analysis in 2-D of Tokamak Hybrid Shielding," IEEE Symp. on Engineering Problems of Fusion Reactors, San Francisco (Nov. 1979).
17. K. D. Lathrop, "DTF-IV, A FORTRAN IV Program for Solving the Multigroup Transport Equation with Anisotropic Scattering," LA-3373, Los Alamos Scientific Lab. (1965).
18. Joseph J. Prabulos, Jr., "DTF-BURN: A Multigroup Transport Theory Burnup Program Written in FORTRAN-IV," LA-3671, Los Alamos Scientific Lab. (1967).
19. D. P. Satkus and H. P. Flatt, "SIZZLE: A Fast or Intermediate Reactor Burnup Code," AI-640Memo-183, Atomics International, Canoga Park, CA (1964).
20. W. W. Engle, Jr., "A User's Manual for ANISN, A One-Dimensional Discrete Ordinates Transport Code with Anisotropic Scattering," K-1693, Computing Technology Center, Oak Ridge National Lab. (1970).
21. R. G. Soltesz and R. K. Disney, "One-Dimensional, Discrete Ordinates Transport Technique," A Description of the ANISN-W Code, Vol. 4 of: "Nuclear Rocket Shielding Methods, Modification, Updating, and Input Data Preparation," WANL-PR-(LL)-034, Westinghouse Astronuclear Lab., Pittsburgh PA (1970).
22. H. A. Sandmeier, G. H. Hansen, R. B. Lazarus, and R. J. Howerton, "Computation of Fast Neutron Penetration in Air by the S_n Method with Special Emphasis on the Use of Multitable-Multigroup Cross Section Sets," LA-3415, Los Alamos Scientific Lab. (1965).
23. G. R. Handley and C. M. Hopper, "Validation Checks of the ANISN and KENO Codes by Correlation with Experimental Data," Y-1858, Union Carbide, Oak Ridge TN (1972).
24. R. K. Disney, J. Jedruch, R. G. Soltesz, R. S. Kaiser, and S. L. Zeigler, "Annual Report for the Contract Year 1970, Supporting Research and Technology, Project 493," WANL-TME-2778, Westinghouse Astronuclear Lab., Pittsburgh, PA (1971).
25. S. A. Dupree, H. A. Sandmeier, G. E. Hansen, W. W. Engle, Jr., and F. R. Mynatt, "Time-Dependent Neutron and Photon Calculations Using the Method of Discrete Ordinates," LA-4557, Los Alamos Scientific Lab. (May 1971).

6.2. MULTIDIMENSIONAL TRANSPORT THEORY CODES

The Oak Ridge DOT Code

The DOT Code[1] was developed by W. A. Rhoades and Fred R. Mynatt of Oak Ridge National Laboratory for solution, by the method of discrete ordinates in the S_n approximation, of the neutron and photon transport equation in two dimensions. The theory for the code was developed by Mynatt as part of his work for a Ph.D. degree.[2] He also wrote the user's manual for the code.[3] Similarly to ANISN, several versions of the code are available, with modifications and improvements made by different laboratories.[4]–[5]

The geometries available for the solution of directed fluxes are (x,y), (r,z) and (r,θ). Anisotropic scattering is permitted between all energy groups. In addition, a diffusion theory solution may be specified for any or all of the energy groups. In this mode of solution the direction and the starting point in the mesh sweep for the flux solution may be chosen in a number of ways, a distinct advantage in complex problem geometries or off-center source locations.

The types of problems that can be solved by DOT are:

- Fixed source, volume distributed or placed at a boundary
- Eigenvalue calculation
- Time absorption (Rossi alpha)
- Material concentration search
- Zone thickness search

The present version of the code provides for variable dimensioning with regard to the number of:

- Energy groups
- Angular components of cross sections
- Direction components of computed fluxes
- Isotopes and mixtures
- Space mesh points and regions

In consequence of variable dimensioning of the computed arrays, only the size of the random access memory of the computer and physical considerations of the problem limit the choice of values assigned to the above variables.

Similarly to ANISN, coupled neutron–photon problems can be solved simultaneously, provided suitable coupled multigroup, multitable cross-section sets are input.

Five schemes for computing mesh fluxes are available—the first three without negative flux fix up, the last two with it:

- Step function
- Weighted difference
- Linear diamond difference
- Linear diamond difference with step function flux fix up
- Linear diamond difference with weighted difference flux fix up

In addition, activity production rates can be computed for reactions for which cross sections have been added to the cross-section table in the extra head positions.

A variety of codes for processing DOT output are available to compute radiation heating rates, dose rates, flux plotting,[6] coupling of leakage fluxes to point kernel calculations,[7] etc.

TWOTRAN, THREETRAN and TRIDENT Codes of Los Alamos

The TWOTRAN Code,[8]–[9] developed at the Los Alamos Scientific Laboratory by D. K. Lathrop and F. W. Brinkley, and the THREETRAN Code,[10] developed subsequently by Lathrop, are discrete ordinate transport theory codes capable of solving the Boltzmann transport equation in the S_n approximation in two and three dimensions. The two codes share many features, although the present version of the THREETRAN code is limited to the rectangular orthogonal (x,y,z) system, whereas TWOTRAN will handle (x,y), (r,z) and (r,θ) geometries. The codes use the central difference scheme in conjunction with the standard finite difference approximation to the spatial derivatives of the discrete ordinate streaming operator.

The inner iterations are done on the scattering source, and are accelerated by updating the scalar fluxes, after the angular fluxes in each quadrant are computed, and by a special line rebalance. The outer iterations are carried out on the fission source, and are unaccelerated in the THREETRAN code. In the TWOTRAN code the inner and outer iterations are accelerated by rebalancing on a coarse mesh which may be independent of the material mesh. The flux solution is obtained subject to vacuum, reflective, periodic, white or input specified boundary flux conditions.

In its present form the THREETRAN code, although fully operational, is still considered to be in a preliminary, unoptimized form, and many service functions performed currently by the one- or two-dimensional codes are not included. There is no cross-section processing in the input or output of the code, and all cross sections must be entered in macroscopic form. Similarly, only one geometry treatment (x,y,z) is at present available.

The current capability of both codes includes:

- Inhomogeneous source
- Eigenvalue problem
- Mixed fixed source and fission source

In addition, TWOTRAN can perform eigenvalue searches, and allows general anisotropic scattering, and anisotropic inhomogeneous sources. The principal limitations of the THREETRAN code are as follows:

- Isotropic scattering only is allowed, although linearly anisotropic scattering could be used without an increase in storage requirements by eliminating the scalar flux update.
- No eigenvalue searches are possible.
- There is no provision for upscattering cross sections.

A TWOTRAN problem in six energy groups, S_4, with 42 by 42 mesh points requires about 3.9 minutes of CDC-7600 time for simple eigenvalue calculation.

The THREETRAN code needs a large computer configuration for implementation, with up to ten disk units required, depending on the problem type and output options. A six-group, S_4, 14,502 mesh problem run requires 144 minutes on the CDC-7600 computer to converge the eigenvalue to 5×10^{-4}.

The TRIDENT code[11] for discrete ordinate transport calculations was developed by T. J. Seed, W. F. Miller and F. W. Brinkley of Los Alamos Laboratory as a modification of the TWOTRAN code.[9] The code solves the Boltzmann transport equation using the S_n approximation in (x,y) and (r,z) geometries modeled in a regular triangular mesh.

The code is particularly suitable for situations involving nonorthogonal material boundaries where ray effects are present, or where oscillation in the flux solution, attributable to the diamond difference calculation scheme, causes problems in convergence. The elimination of ray effects is accomplished by appending a fictitious source to the discrete ordinate equations, which is defined such that in the (x,y) geometry the P_{n-1} spherical harmonic solution or in the (r,z) geometry the DH_{n-1} spherical harmonics-like solutions are obtained.

The application of the nonorthogonal boundaries in the solution reduces to the order of the square root of 2 the error in the value of neutron currents computed for any curvilinear interface, as compared to the results of the same calculation setup in orthogonal mesh representation. This consideration is particularly important in reactors loaded with hexagonal assemblies or representing complex geometries, such as occur in toroidal fusion reactors.

The code allows general anisotropic scattering and will accept anisotropically distributed sources.

In the solution for directed fluxes, both inner (within group) and outer (source) iteration cycles are accelerated by whole-system or fine-mesh rebalance.

The major features of the code include:

- Direct or adjoint flux calculation option.
- Built-in S_n constants.
- Option to input quadrature sets.
- Recurring, reflective or source boundary conditions.
- Nonrectangular domain boundaries.
- Source, eigenvalue, Rossi-alpha, or concentration searches.
- Print suppression for large input arrays.
- Space-dependent buckling.
- Optional input flux, source, S_n constants and cross sections from standard interface files.
- Optional output of scalar and angular fluxes and S_n constants in standard interface format.
- Flexible data management algorithms that allow automatic overflow to a random disk for large problems, if necessary.

The constraints of geometry handled by the code are:

- Adjacent triangles in the geometry setup of the problem must share a common vertex.
- All triangles must lie on horizontal bands extending across the system.
- Each interior vertex must be common to exactly six adjacent triangles.

A new version(12) of TRIDENT has been developed recently to handle the specific needs of the fusion community. It is particularly suited for modeling the toroidal geometry of various tokamak devices.

REFERENCES

1. W. A. Rhoades, D. B. Simpson, R. L. Childs, and W. W. Engle, Jr., "The DOT-IV Two-Dimensional Discrete Ordinates Transport Code with Space-Dependent Mesh and Quadrature," ORNL/TM-6529, Oak Ridge National Lab. (1979).

2. Fred R. Mynatt, "Development of Two-Dimensional Discrete Ordinates Transport Theory for Radiation Shielding," Doctoral Thesis, University of Tennessee, Knoxville TN (1969).
3. F. R. Mynatt, "A User's Manual for DOT, A Two-Dimensional Discrete-Ordinates Transport Code with Anisotropic Scattering," informal notes included in RSIC Computer Code Collection, CCC-89, Oak Ridge National Lab. (1969).
4. R. G. Soltesz, R. K. Disney, J. Jedruch, and S. L. Zeigler, "Two-Dimensional, Discrete Ordinates Transport Techniques, A Description of DOT-IIW, DOQ, ADOQ, and MAP Codes," Nuclear Rocket Shielding Methods, Modification, Updating and Input Data Preparation, Vol. 5, WANL-PR-(LL)-034, Westinghouse Astronuclear Lab. (1970).
5. R. G. Soltesz, R. K. Disney, J. Jedruch, S. L. Zeigler, and G. Collier, "User's Manual for the DOT-IIW Discrete Ordinate Transport Computer Code," WANL-TME-1982, Rev. A, Westinghouse Astronuclear Lab. (1970).
6. L. L. Moran and R. G. Soltesz, "FLUXPLT, A FORTRAN-IV Program to Plot Isoflux Lines Based on the Scalar Flux Calculated by the DOT Program," WANL-TME-1813, Westinghouse Astronuclear Lab. (1968).
7. R. K. Disney, R. G. Soltesz, J. Jedruch, and S. L. Zeigler, "Code Description and User's Manual for the MAP Radiation Transport Code," WANL-TME-2706, Westinghouse Astronuclear Lab. (1970).
8. K. D. Lathrop and F. W. Brinkley, "Theory and Use of the General-Geometry TWOTRAN Program," LA-4432, Los Alamos Scientific Lab. (1970).
9. K. D. Lathrop and F. W. Brinkley, "TWOTRAN-II: An Interfaced Exportable Version of the TWOTRAN Code for Two-Dimensional Transport," LA-4848-MS, Los Alamos Scientific Lab. (1973).
10. K. D. Lathrop, "THREETRAN: A Program to Solve the Multigroup Discrete Ordinates Transport Equation in (x,y,z) Geometry," LA-6333-MS, Los Alamos Scientific Lab. (1976).
11. T. J. Seed, W. F. Miller, and F. W. Brinkley, "TRIDENT—A Two-Dimensional Triangular Mesh Discrete Ordinates, Explicit Neutron Transport Code," LA-6735-M, Los Alamos Scientific Lab. (1977).
12. Thomas J. Seed, "TRIDENT-CTR User's Manual," LA-7835-M, Los Alamos Scientific Lab. (1979).

6.3 DIFFUSION THEORY WITH BURNUP CODES

The Hanford 2DB and 3DB Codes

The 2DB and 3DB Codes[1]–[3] were developed by W. W. Little and R. W. Hardie of Battelle Northwest Laboratories for the calculation of two- and three-dimensional fuel burnup and criticality in the multigroup diffusion theory approximation. The 2DB Code handles (x,y), (r,z) and (r,θ) geometries, whereas 3DB will solve the criticality problems with burnup in three dimensions with (x,y,z), (r,z,θ) or (triangular,z) geometries. Due to their efficient handling of multigroup energy structure, which allows relatively detailed representation of the neutron spectrum at all energies, the codes are particularly suitable for burnup analysis of fast breeder reactors.

One of the features of the codes is variable dimensioning of the computed arrays, so that in a problem setup complexity in the energy group structure may be traded off against detail in the space mesh definition. In computers with 65 kilowords of random access memory, the current version of the 2DB code will handle up to 50 energy groups. However, since in both 2DB and 3DB only one energy group and in 3DB only one plane of geometry are in the fast memory at any given time, the computer storage requirements are relatively insensitive to the number of energy groups and the number of space intervals.

In the solution of the diffusion part of the problem the multigroup energy mesh is used, and the space dependent multigroup fluxes are then used to collapse the cross sections for the burnup calculation at each space point. Thus, in the depletion calculation, effectively space- and energy-dependent cross sections are used. This feature makes it possible for the calculation to account for spectral variations present near interfaces between the core and the breeding blankets, as well as between the blankets and the surrounding cooling medium in the breeder reactors.

The eigenvalue is computed by the code using the standard source iteration techniques. Group rebalancing and pointwise overrelaxation methods are used to accelerate convergence. Adjoint flux solutions are obtained by inverting the input data and redefining the source terms.

The codes have capability for performing criticality searches on:

- Buckling
- Time absorption alpha
- Material concentrations
- Region dimensions

In both codes criticality searches can be performed during burnup, to compensate for fuel depletion. Alpha and k_{eff} can be used for parametric eigenvalues. If required, the codes will compute the effective multiplication constant and the flux profiles corresponding to an arbitrary specification (in space and energy) of an extraneous neutron source.

The format of the input data, i.e., cross sections, flux guess, geometry and composition specifications, is compatible with Los Alamos one- and two-dimensional transport theory codes DIF-IV and 2DF, uses the same input module, and differs from the DOT code input merely in the names assigned to the input arrays. In fact, the latest version of the DOT code includes the 2DB diffusion solution module to generate the first iteration scalar flux as a first guess for the transport solution.

Both codes require 65 kilowords of space in the random access memory of the computer. In addition, the 3DB code requires 11 peripheral storage units.

The running time of the 2DB code for a k_{eff} calculation is only about 2 minutes on UNIVAC-1108 for a 30 × 30 mesh point problem when run in four energy groups for one time step. The running time of the 3DB code on the same machine is about 0.5 hour for the first time step for a 20 × 20 × 20 mesh problem in two energy groups. Each succeeding time step takes about a quarter of the above time.

These codes are representative of several codes of this type and are user-friendly by clear formatting and checking of input and output, as well as providing output in a form suitable for direct use in fuel cycle engineering. The output of the codes includes tabulations of heavy metal and nondepletable material inventories by zone, region and time step, as well as the power generated in these regions.

Bettis PDQ Code

The PDQ Code[4] was written in its first version by G. G. Bilodeau, W. R. Cadwell, J. P. Dorsey, J. G. Fairey and R. Varga of the Bettis Atomic Power Laboratory to solve the neutron diffusion equations with depletion in one, two or three dimensions. Subsequent versions were written by W. R. Cadwell,[5] and the latest version by C. J. Pfifer and C. J. Spitz.[6] Versions of the code exist for all major computers. This code is one of the principal design tools used in reactor analysis and has been in use for almost thirty years.

The code is capable of accommodating very large problems, solved in five energy groups. The thermal energy group can be represented by a single group, or by two overlapping groups. The spatial solution for the flux may be done in an explicit mode, or discontinuous trial functions may be used with flux synthesis. This method was developed mainly at the Bettis Laboratory, and is described in Chapter 11 of Alan Henry's *Nuclear Reactor Analysis*.[7]

The group equations are solved in the code directly, in one dimension using Gauss elimination, switching to the single-line cyclic Chebishev semi-iterative method in two dimensions. In three dimensions, the single-line successive overrelaxation method is used. The calculation of source iterations is accelerated by the application of the Chebishev extrapolation to group fluxes.

The following types of calculations may be performed:

- Eigenvalue
- Boundary value
- Fixed source, or additive fixed source
- One iteration
- Adjoint problem

The depletion part of the calculation is based on the HARMONY[8] system of equations, which allows the time-dependent representation of cross sections, and uses a generalized representation of the depletion chains.

The geometry treatment of the code allows variable mesh in all dimensions and geometries. Rectangular, cylindrical, spherical and hexagonal geometries are available. In addition to the above features, temperature and xenon feedbacks are provided in the code.

All data storage in the code is dynamic, and the maximum allowed product of energy groups and space mesh points is approximately ten million, due to the limit on the magnetic disk capacity in CDC-7600 computers. CDC-6600 storage capacities are about an order of magnitude smaller. The running times in hours on CDC-6600 are estimated by dividing the product of groups and mesh points by 150,000.

Considering that version one[4] of PDQ came out nearly thirty years ago (1957) its eighth version,[6] available since 1978, makes it the true veteran of the field.

The Argonne DIF3D and DIF3D (NODAL) Codes

The DIF3D and DIF3D (NODAL) codes[9]–[11] for the solution of multidimensional neutron diffusion equation were written respectively by K. L. Derstine and R. D. Lawrence of Argonne National Laboratory. In addition to the solution for the eigenvalue, the DIF3D code treats the adjoint problem, the fixed source problem and criticality search problems. Orthogonal rectangular, cylindrical or triangular one-, two- or three-dimensional geometries can be used in the solution. The DIF3D (NODAL) version of the code handles only two- and three-dimensional problems in hexagonal geometry. Both codes will accept anisotropic diffusion coefficients. Although the codes are primarily intended for solving fast reactor problems, they can also be used for reactors supporting soft energy spectra, since upscattering and internal black boundary conditions can be treated.

In the DIF3D calculation, the mesh-centered finite difference equations are solved by iteration techniques optimized by the addition of Chebishev acceleration applied to outer fission source iterations, and the block successive overrelaxation method applied to the inner group iterations. In nonperiodic orthogonal geometry cases, prior to the initiation of the outer iteration, the decomposition algorithm for the resulting tri-diagonal matrices is applied. After convergence is reached, the maps of flux and power are computed by mesh cell, together with regionwise balance integrals.

In the nodal version of DIF3D, hexagonal geometry only is allowed, and a single meshpoint per hexagon is used instead of the usual six triangular meshpoints typically employed in fast reactor finite difference calculations. In the axial direction, higher order approximation is used with coarse mesh without sacrificing accuracy. The nodal equations in the solution of the diffusion equation are derived using higher order polynomial approximations of the spatial dependence of the flux within the hexagonal node. Nodal coupling coefficients are precomputed and stored only for unique nodes. The final equations are cast in response matrix form, and involve spatial moments of the node interior flux distribution plus surface averaged partial currents across the faces of the node.

Both versions of DIF3D read and write standard interface files specified by the Committee on Computer Code Coordination (CCCC). Their burnup function is supplied by the REBUS Code,[12] which uses the fluxes computed by DIF3D in its depletion calculations.

Both codes are implemented on the IBM 370/195 and the CDC-7600 machines, and have been tested on the CRAY-1 machine in a nonvectorized version. Both versions of the code require at least 300k bytes of core storage for program and file buffer storage on the IBM 370 machine. On the CDC-7600 machine between 20 and 30k words of space are required in the Small Core Memory depending on the operation system used. DIF3D requires additional Large Core Memory space that expands linearly with the number of mesh cells in a mesh plane, and at least 19 mesh lines of data and one group of cross section data must be stored in the Small Core Memory. For both codes the external data storage requires approximately 40 scratch and interface files.

The running time of DIF3D is measured in flux work units, equal to the product of space mesh cells times the number of energy groups times the number of iterations per group. In three dimensional problems, 4 to 8 million flux work units can be computed per minute by the IBM 370/195, with slightly shorter run times on the CDC-7600, and one-third the run times on the CRAY-1. The run times of the DIF3D (NODAL) version are substantially smaller. A three-dimensional calculation in 8 energy groups, 17 axial mesh planes and 17 rings of hexagons, when treated in the 1/6 core symmetry model requires 2 CPU minutes for solution by the IBM 370/195 machine.

REFERENCES

1. W. W. Little and R. W. Hardie, "2DB, A Two-Dimensional Diffusion-Burnup Code for Fast Reactor Analysis," BNWL-640, Pacific North-West Lab., Richland, WA (1968).
2. W. W. Little, R. W. Hardie, "2DB User's Manual," BNWL-831 Pacific Northwest Lab., Richland WA (1968).
3. R. W. Hardie, W. W. Little, "3DB, A Three-Dimensional Diffusion Theory Burnup Code," BNWL-1264, Battelle Memorial Institute, Pacific Northwest Lab., Richland WA (1970).

4. G. G. Bilodeau, W. R. Cadwell, J. P. Dorsey, J. G. Fairey, and R. S. Varga, "PDQ—An IBM-704 Code to Solve the Two-Dimensional Few-Group Neutron Diffusion Equations," WAPD-TM-70, Bettis Atomic Power Lab. (1957).
5. W. R. Cadwell, "PDQ-7 Reference Manual," WAPD-TM-678, Westinghouse Electric Corp. (1967).
6. C. J. Pfeifer and C. J. Spitz,"PDQ08 Reference Manual," WAPD-TM-1266, Bettis Atomic Power Lab., Pittsburgh PA (1978).
7. Alan Henry, *Nuclear Reactor Analysis,* The MIT Press, Cambridge, MA (1975).
8. R. J. Breen, O. J. Marlowe, and C. J. Pfeifer, "HARMONY: System for Nuclear Reactor Depletion Calculations," WAPD-TM-478, Bettis Atomic Power Lab. (1965).
9. K. L. Derstine, "DIF3D: A Code to Solve One-, Two-, and Three-Dimensional Finite Difference Diffusion Theory Codes," ANL-82-64, Argonne National Lab. (1983).
10. D. R. Ferguson and K. L. Derstine, "Optimized Iteration Strategies and Data Management Considerations for Fast Reactor Finite-Difference Diffusion Theory Codes," *Nucl. Sci. Eng.*, 64, 593 (1977).
11. R. D. Lawrence, "The DIF3D Nodal Neutronics Options for Two- and Three-Dimensional Diffusion Theory Calculations in Hexagonal Geometry," ANL-83-1, Argonne National Lab. (Apr. 1983).
12. B. J. Toppel, "The Fuel Cycle Analysis Capability, REBUS-3," ANL-83-2, Argonne National Lab. (Mar. 1983).

6.4 STOCHASTIC METHODS

Early in the Manhattan Project it became obvious to the participating analysts that the essentially random nature of nuclear processes could be mathematically simulated on an electronic computer, giving the reactor physicist and the nuclear engineer better insight into the actual processes. The success of this approach required that two conditions be fulfilled: a satisfactory random number generator had to be developed, and computing machinery capable of following tens of thousands of life histories of relevant processes had to be designed. Both of these objectives were accomplished, and the method was put to work.

The first papers[(1),(2)] defining the digital applications of the Monte Carlo method were authored by Stanislaw Ulam in cooperation with Nicholas C. Metropolis, and appeared in the late 1940s. By the mid-1950s the method was being applied to a variety of problems in reactor physics, such as resonance capture, particle transport, computation of thermal utilization factors, etc., adding to the store of available computer programs.[(3)–(5)] The earliest engineering applications of the method are due to the Knolls[(6)] and the Bettis Atomic Power Laboratories with the TRAC and RECAP[(7),(8)] codes developed in the late 1950s and the early 1960s. In the late 1960s the Monte Carlo method came into its own; the 1970s saw the proliferation of its applications and the subsequent development of the method into a computing tool[(9)] of great versatility and power.

The trends discernible in the development of the Monte Carlo computer codes are: the desire to reduce computing time and computer memory storage requirements, the introduction of sophisticated techniques for killing particle life histories in regions of low interest through so-called games of Russian roulette, the multiplication of life histories in regions of high interest through particle splitting, and, finally, the introduction of elaborate geometrical and tally routines. When these analytical advances were combined with advances in the development of computer hardware, the Monte Carlo technique was removed from the realm of infrequently used methods of last recourse, and entered into routine use in engineering analysis.

Oak Ridge MORSE and KENO-V Codes

The MORSE Code[10] was developed by E. A. Straker, P. N. Stevens, D. C. Irving and V. R. Cain of Oak Ridge National Laboratory as a tool for studying neutron and gamma-ray transport by stochastic methods. A new version of the code, MORSE-SGC, has improved running capability and smaller computer storage requirements. Both versions are capable of treating the transport of coupled neutron–gamma-ray fluxes in direct or adjoint modes.

The main version of MORSE accepts coupled neutron–photon multi-group cross-section libraries in the DTF-IV, ANISN or DOT format. The MORSE-SGC version accepts cross sections in the AMPX working library format.[11] A subsidiary code, LAVA, is available to convert cross sections from the ANISN format used by MORSE to the AMPX format used by MORSE-SGC. The MORSE-SGC version is available with either the combinatorial geometry package or the KENO geometry package.

With a modular input–output structure, the code also has:

- An output routine that handles conversion of random-walk events into reaction rates.
- Input debugging routines.
- Time-dependence capability for shielding and criticality problems.
- The albedo option[12] at any material boundary.
- One-, two-, and three-dimensional geometry packages.
- Several types of options for importance sampling.

The time-dependent capability of the code makes it particularly suitable for analyses of the inertially confined fusion–fission hybrid reactors, or for treating fission reactor kinetics in the prompt critical region.

The MORSE-SGC version of the code reduces computer storage needs for problems with large cross-section requirements. This is accomplished by storing the cross-section data on direct access devices, and retrieving super groups of cross-section data (blocks consisting of one or more energy groups) as needed by the code. Supergroup sizes are calculated by the program to fit within the random access core size allocated.

The KENO-V code[13] is a multigroup, combinatorial geometry Monte Carlo criticality code. Its primary function is to compute multiplication constants of nuclear materials arranged in a three-dimensional configuration. The code was conceived and originally developed by G. E. Whitesides and N. F. Cross of Oak Ridge National Laboratory in 1969.[14] Since its first appearance, the code has enjoyed unflagging popularity with analysts, has been updated and improved several times by H. K. Clark, L. M. Petrie and N. F. Landers, and has just become available in version five.

The quantities computed by KENO include k_{eff} of the system, with an estimate of its standard deviation, plus a tally of leakage, absorptions and fissions in each energy group. A fission probability matrix is computed, to represent interactions within the system for each array position, unit number, hole or array number. Average self multiplication of the fissioning units in the array, neutron lifetime and generation time are also computed. The k_{eff} of the system is computed by solving the homogeneous integral Boltzmann equation, or is obtained optionally as the largest eigenvalue of the interaction matrix.

The special geometry package included with the code allows complex geometries to be constructed from any combination of cylinders, spheres and rectangular parallelepipeds. "An array of arrays" can also be built within the same problem from arrays of different size and shape, with the constraint that adjacent faces of units within one array must match exactly in size and shape. An additional "hole" capability of the geometry allows placement of regions of differing properties within a larger region, provided they do not intersect. This capability is particularly useful for representing fuel shipping casks, reflectors with gaps in them, or control rods embedded in an array.

In the KENO method of solution the differential scattering cross sections are represented by a P_1 Legendre polynomial. The absorption of neutrons is represented in KENO by reducing the weight of the neutron at each collision point by its absorption probability. When the weight of the neutron has been reduced beyond a certain point, its further fate is decided in a game of Russian roulette. If it survives, its history is continued with an increased weight. In order to minimize the variance in k_{eff}, splitting of high-weight neutrons is carried out, when regions of widely varying weights are present.

The cross-section libraries required by KENO must be in a format compatible with either the AMPX system or the format generated for Monte Carlo calculations by the ICE-s program. The code has been used with a 16-group Hansen–Roach cross section set, a 27-group ENDF IV set, a 123-group GAM–THERMOS set and a 218-group ENDF IV set. Albedos for reflector materials with a small absorption-to-scattering ratio can be used in the form of input files.

The energy-dependent data, such as cross sections, albedos, weights, leakages, absorptions fissions and fluxes, are handled in supergroups of size chosen internally by the computer. Thus, even large problems can be run on a small computer, since only one supergroup is residing in the computer memory at any given time.

Version IV of KENO[15] has been implemented on both the CDC-7600 and the IBM-360 machines. The k_{eff} of an array of 2 by 2 by 2 of highly enriched cylinders can be obtained with a standard deviation of 0.52% from 22,500 neutron life histories in 0.281 minute of computing time on the IBM-360/195. With a reflector of 6 inches of paraffin, 18,600 life histories gave a value of k_{eff} with 0.69% standard deviation in 2.249 minutes of computing time on the same machine.

Version V of KENO is operational on the IBM 3033 and IBM 360/195 machines. On the first of them 1/2% standard deviation in k_{eff} of a simple unreflected, unmoderated assembly can be obtained in less than 0.5 minute. A running time of 30 minutes can be expected for the same standard deviation of a complicated array with reflector.

Los Alamos MCNP Code

The MCNP Code[16] was developed as a collective effort of Group TD-6 at the Los Alamos National Laboratory for the computation of static and time-dependent neutron and photon transport by Monte Carlo methods. Its capabilities include the treatment of geometric shapes such as the plane, cylinder, sphere, cone, ellipsoid, hyperboloid, paraboloid, torus or a combination thereof, provided they can be described by first-, second-, or fourth degree surfaces.

The treatment of energy dependence of cross sections in MCNP permits the inclusion of all reactions that are specified in the ENDF/B file at each energy point. The reactions considered in treating electromagnetic radiation include coherent and incoherent scattering, photoelectric absorption followed by fluorescent emission, pair production and coincident annihilation radiation.

Among the options available for variance reduction are the following:

- Geometry splitting
- Russian roulette
- Exponential transformation

- Energy splitting
- Forced collisions in designated cells
- Flux estimates at a point detector
- Track length estimators
- Source biasing

The output of the code produces the following quantities:

- Two-way current is obtained as a function of energy, time and angle with the normal directions across any set of bounding surfaces.
- Fluxes, i.e., total track length per unit volume, as a function of energy and time, are available at any bounding surface or selected point, or as an average in the cell.
- Reaction rates, such as fissions or absorptions, may be obtained in any subset of geometric cells.
- Energy depositions per starting particle can be tallied.
- Particles can be flagged as they cross designated surfaces, and their contribution to certain reactions can be listed.
- Errors in all quantities calculated are also listed.

This code is available for use at the Magnetic Fusion Energy (MFECC) Computer Center at the Lawrence Livermore Laboratory for those laboratories that have access to MFE by its cable network.

Lawrence Livermore TARTNP Code

The TARTNP Code[17] was developed by E. F. Plechaty and J. R. Kimlinger at the Lawrence Livermore Laboratory for the study of neutron and photon transport by stochastic methods. The code was created by merging the TART neutron transport code and the TORTE photon transport code. Transport of photons generated by neutron-induced reactions is included in the analysis. Particle reflections from planes and quadratic surfaces, including spectral reflection, are available in the code options. The code can be run in either the reaction probability mode or the emission probability mode.

The cross sections for particle transport used by the code are derived from the Livermore ENDL library, and cover the energy range from 20 MeV down to 1 keV for photons, and 1 meV for neutrons.

In setting up the geometry of calculation, the user can divide the problem into a maximum of 110 zones, and can choose any one of 17 methods of scoring particle transport (output tallies). A tally of neutron and photon energy deposition by zone is included in the output options. Zones to be examined, and elements to be included in the tally, can be specified by the user.

The kinematic methods available in the code refer to the choice of ways to describe the speed and the scattering angle of the transported neutrons and photons after a collision event. The representative kinematic methods include:

- Scattering, elastic and inelastic, to a level with a Q-value.
- Treatment of wide energy levels described by the single-level dispersion theory.
- Sampling of tabular distributions in either angle or energy at specific energies.
- Treatment of special cases, such as the (n,2n) reactions assumed to be a time-sequential process, in which the second neutron travels at the same speed and in the same direction as the recoiling nucleus and is not bound to it.
- A combination model, which combines the discrete and the wide-level methods with the uncorrelated tabular spectral treatment.

Four methods for calculating the energy and emission angle of a photon are available:

- Isotropic monoenergetic photon
- Anisotropic monoenergetic photon
- Isotropic with tabular energy distribution
- Anisotropic with tabular energy distribution

Limitations of the code include a lack of provisions for correlated energy angular distributions.

SAM-CE MAGI Code

The SAM-CE Code[18] was developed by M. O. Cohen, E. Troubetzkoy, M. Beer and H. Lichtenstein of the Mathematical Applications Group, Inc., to analyze by stochastic methods the neutron and gamma-ray transport problem, as well as the transport of neutron-induced secondary gamma rays. Both forward and adjoint calculations can be implemented.

Time-dependent fluxes and flux-related functions such as radiation dose, heatup rates, etc., may be calculated as a function of energy, time and position. Collision and absorption densities can be scored over either finite volume elements or selected regions.

A combinatorial geometry treatment offers a capability for constructing complex geometrical regions. When complex shapes are constructed by intersecting simpler shapes, duplication of volumes in the overlapping regions is prevented by an automatic geometry checker which forms part of the code.

The code uses nuclear cross sections derived from the ENDF-IV data file, and tabulates neutron cross section values on an energy mesh internally derived on the basis of built-in convergence criteria and tolerances specified by the user. In this scheme, each nuclide has associated with it a unique energy mesh appropriate for it. Extremely precise description of the cross-section behavior is possible, with resolved and unresolved resonance parameters from the ENDF data file treated automatically. Such treatment overcomes the difficulty present with the multigroup codes using flux-averaged cross sections based on assumed flux distributions, which may not adequately represent the real flux distributions in a given region.

By splitting the cross-section data into two or more energy ranges, to be treated one at a time, the code makes it possible to consider many nuclides in much detail in a given configuration. Energy, region and angular importance sampling are possible.

The code is organized into three separate programs:

- SAM-X preprocessor for neutron and gamma-ray cross sections and gamma-ray production data.
- SAM-F for forward calculation.
- SAM-A for adjoint calculation.

In SAM-X the following reaction types are treated as components of the absorption cross section: (n,γ), (n,p), (n,d), (n,t), (n,He^3), (n,α), $(n,2\alpha)$, $(n,3\alpha)$, (n,f).

The following reaction types are treated as components of the inelastic cross section: (n,n'), (n,2n), (n,3n), $(n,n)\alpha$, $(n,2n)\alpha$, $(n,3n)\alpha$, (n,n')p, $(n,n')3\alpha$.

REFERENCES

1. N. C. Metropolis and S. M. Ulam, "Monte Carlo Method," *J. Am. Statist. Assoc.*, 44, 335–341 (1949).
2. "Publications of Stanislaw M. Ulam," compiled by Barbara Hendry, LA-3923-MS, Los Alamos Scientific Lab. (1968).
3. R. D. Richtmeyer, "Monte Carlo Study of Resonance Escape in Hexagonal Reactor Lattices," NYO-6479 (1955).
4. H. D. Brown, "A Monte Carlo Study of Neutron Thermalization," AECU-3758 (1957).

5. M. H. Kalos, "A Monte Carlo Calculation of the Transport of Gamma Rays," NDA-56-7 (1956).
6. M. A. Martino and W. W. Stone, "A Monte Carlo Thermal Neutron Code for the IBM-704," KAPL-2039 (1959).
7. H. J. Berwind and J. Spanier, "TRAC-1, A Monte Carlo Philco-2000 Program for the Calculation of Neutron Capture Probabilities," WAPD-TM-229 (1961).
8. N. R. Candelore and R. C. Gast, "RECAP1—A Monte Carlo Program for Estimating Epithermal Capture Rates in Slabs," WAPD-TM-247 (1963).
9. J. Spanier and E. M. Gelbard, *Monte Carlo Principles and Neutron Transport Problems,* Addison-Wesley Publishing Co., Reading, MA (1965).
10. E. A. Straker, P. N. Stevens, D. C. Irving, and V. R. Cain, "The MORSE Code: A Multigroup Neutron and Gamma-Ray Monte Carlo Transport Code," ORNL-4585, Oak Ridge National Lab. (1970).
11. "AMPX-II, A Modular Code System for Generating Coupled Multigroup Neutron-Gamma Libraries from ENDF/B, PSR-63 Package," Radiation Shielding Information Center, Oak Ridge National Lab. (1978).
12. R. E. Maerker, "The Use of the Albedo Option in MORSE," ORNL-RSIC-29, pp. 125–132 (1971).
13. N. F. Landers and L. M. Petrie, "KENO V.a—An Improved Monte Carlo Criticality Program with Extended Geometry Capabilities and Supergrouping," Vol. 1, pp. 214–217, "Advances in Reactor Computations," *Proc.* ANS Topical Meeting, Salt Lake City (Mar. 28–31, 1983).
14. G. E. Whitesides and N. F. Cross, "KENO—A Multigroup Monte Carlo Criticality Program," CTC-5 (Sept. 1969).
15. L. M. Petrie and N. F. Cross, "KENO-IV—An Improved Monte Carlo Criticality Program," ORNL-4938, Oak Ridge National Lab. (1975).
16. Los Alamos Monte Carlo Group, "MCNP—A General Monte Carlo Code for Neutron and Photon Transport," LA-7396-M, rev., Los Alamos National Lab. (1981).
17. E. F. Plechaty and J. R. Kimlinger, "TARTNP: A Coupled Neutron–Photon Monte Carlo Transport Code," UCRL-50400, Vol. 14, Lawrence Livermore Lab. (1976).
18. M. O. Cohen, H. Steinberg, E. Troubetzkoy, M. Beer, and H. Lichtenstein, "SAM-CE: A Three-Dimensional Monte Carlo Code for the Solution of the Forward Neutron and Forward and Adjoint Gamma Ray Transport Equations—Revision C," DNA-2830F (1974).

7. STRUCTURAL, THERMO-HYDRAULIC AND SAFETY ANALYSIS

7.1 FINITE ELEMENT CODES FOR STRESS ANALYSIS

The impact of machine methods of analysis on structural and thermo-fluid areas became evident early in the nuclear era. First to feel its influence was stress analysis of vessels and piping, where three-dimensional finite element methods became firmly established and experienced considerable development, primarily under the sponsorship of NASA. Both elastic and plastic deformation analyses were tackled using this method, with creep analysis of heated shapes first performed by Gallagher et al.[1] in 1962, followed by Greenbaum and Rubinstein[2] in 1968. The analytical formulation of the basis of finite element codes is due to O. C. Zienkiewicz,[3] followed by Desai and Abel.[4] Further development of this method was supported by Gallagher, Bathe and Wilson and others,[5]–[10] and recent work in this field is due to Hughes and Cook.[11],[12]

The finite element method of numerical analysis is based on the principle that complex geometrical shapes describing components of structural or mechanical equipment can be completely subdivided into small finite elements by means of an imaginary three-dimensional grid structure. Each of the resulting space elements retains the same physical properties as the larger body, but is assumed to be free to respond to mechanical and thermal perturbations with a specified number of degrees of freedom. The response of each element to the perturbations is subject to constraints imposed on it by its relative position within the larger body. Both solid bodies and fluids can be so treated. Generally, a fine space mesh is required, resulting in the simultaneous solution of very large systems of equations through the application of advanced matrix methods.

Since a large part of the effort required to set up problems for codes based on the finite element formulation goes into the error-free specification of space mesh, early attempts were made to mechanize this part of the analysis. Schemes for mesh optimization were introduced,[10] and all the latest computer codes are equipped with automatic mesh setup and checking.

The specifics of nuclear applications of structural methods are such as to require a capability for predicting stresses, strains and deformations in situations with superposition of mechanical and thermal loading. Thermal loading for reactor vessel components includes radiation heating, often distributed in a complex manner corresponding to the paths of streaming neutrons and gamma rays. In some nuclear applications cyclic thermomechanical loading exists, and the analysis requires the assessment of component integrity throughout plant life, on the order of 30 years. Under certain conditions it is difficult, if not altogether impossible, to carry out periodic inspections of equipment installed in high radiation fields. When this situation is combined with a nonlinear structural response, the ability to verify the quality of analytical predictions carries a high premium.

Subsequent to the formulation of the analytical method of approach, a number of finite element computer codes were developed, taking advantage of the capabilities of computer hardware now available to the analyst. The most powerful of the present-day codes are handle systems with as many as 100,000 degrees of freedom. Codes such as MSC/NASTRAN, ANSYS, MARC-CDC,[15] WECAN[16],[17] and ADINA[18] appeared and found wide applications in the nuclear field. Two examples of the leading codes frequently used for nuclear applications are discussed later to show the range of capabilities of modern codes of this type.

With the analytical tools available through the introduction of the finite element methods, it now became possible to perform thermoelastic plastic and creep analyses of complex shapes associated with inlet and outlet penetrations in reactor vessels, head closures, etc. Other applications of finite element methods are found in calculations of heat transfer in systems dominated by convective, rather than conductive or radiative processes,[11] and in the analysis of problems involving liquid metals.[9]

NASTRAN Code

The NASTRAN Code[19]–[20] was developed by NASA for space applications, including the development of the NERVA rocket engine. At present it is the leading computer code in the area of finite element analysis of structural problems. It is also recognized as one of the most comprehensive programs for finite element analysis in the nuclear field. The latest version of the code in use is designated NASTRAN Level 17.5. Although the code was placed in the public domain in the late 1970s, it exists currently in several proprietary versions featuring extensive enhancements over the original version, the best known being the version supported by the MacNeal-Schwendler Corporation, and designated MSC/NASTRAN.

NASTRAN is a large scale structural code capable of solving both linear and nonlinear analysis problems in diverse applications, not only nuclear but also automotive, shipbuilding, aerospace and others. The range of problems it can handle includes static stress analysis, vibrational analysis, dynamic problems involving structural damping, problems with steady state and transient heat transfer with surface ablation, and several variations of the above.[21] In nuclear applications, MSC/NASTRAN is often used for analysis of reactor pressure vessels and piping, with special emphasis on pressure vessel closures, flanges, penetrations, pipe to pump transition pieces, and the like.

Very large problems, with as many as 100,000 degrees of freedom, can be solved efficiently and accurately with NSC/NASTRAN with heavy dependence on the sparse matrix routines. These routines recognize detailed patterns of nonzero terms, and process them in condensed form. Several methods of matrix multiplication are used, depending on matrix size and density, with efficient automatic spill logic for matrices exceeding in size the random access memory of the computer.

Subdivision of very large problems into superelements is possible to fit the problem into the computer. After separate processing, these superelements are automatically combined to reduce the total to a manageable size. Automatic mesh generation for finite element models can be done with an auxiliary routine, MSGMESH. Cyclic symmetry is recognized and applied to reduce computing time.

To help detect errors in input data describing problem geometry and material characteristics such as elastic properties, mechanical constraints and distribution of loads, automatic input data checking is provided as part of the code. Formats and completeness of input data are verified, and geometrical compatibility of input elements is established before a solution is attempted. Results of input checks are presented in tabular form for the convenience of the analyst.

Built into the code is a large library of more than 50 basic elements such as scalar elements, axisymmetric elements, mass elements, rigid elements, fluid elements and heat transfer elements. User-supplied "dummy" elements can also be used in setting up a problem. Curved pipe, curved shell, rigid shell and solid elements are available in the MSC/NASTRAN. These elements can be combined in setting up large problems for solution. Another convenience feature of the code is the automatic suppression of singularities, available at the user's option. With this feature, nearly-singular or singular degrees of freedom can be suppressed, and an indication of the removed degrees of freedom is provided in the output.

The application of the code is facilitated through the provision of the

Direct Matrix Abstraction Programming (DMAP) to direct NASTRAN in solving general matrix problems. This is a user-oriented language providing macro-instructions for the user to write his own analysis routines. Standard solution sequences are available to conserve effort at setup time.

The code is equipped with an extensive plotting and interactive graphics capability. Orthographic, stereoscopic and perspective projections of all the geometric elements in the code library are available. Undeformed and deformed projections of problems can be plotted. Applied loads can be displayed as vectors. A post-processor, MSGSTRESS, prepares a summary of stress distribution over cross sections of two- or three-dimensional fields previously defined with the MSGMESH data generator.

The NASTRAN program is currently available in versions implemented on the UNIVAC 1100 series, IBM 360/370 and IBM 303x computer series, CDC 6000 and CDC CYBER 70/170 machines and DEC VAX machine series.

Access to MSC/NASTRAN is available under a license from Babcock and Wilcox Computer Services Center, P. O. Box 1260, Lynchburg, VA 24505, tel. (804) 384-5111, or from Control Data Corporation, CYBERNET Services, HQW05H, P. O. Box 0, Minneapolis, MN 55440, tel. (612) 482-2830.

ANSYS Code

The ANSYS Code[(22)–(23)] was developed by P. C. Kohnke in 1977. It is another general-purpose, finite element code designed for structural analysis in one, two or three dimensions, and can handle both static and dynamic problems. Buckling, small and large deflections in the elastic and plastic range can be analyzed, including creep and swelling of the material. These last two options are particularly useful in nuclear applications. Heat transfer analysis can be handled by the code for conduction, convection and radiation in the static and transient domain, and fluid–solid interaction can be handled in coupled thermo-fluid problems.

The code is equipped with an extensive mesh generating capability for handling general two- and three-dimensional intersecting shell geometries. Additional sets of nodal points and elements can be replicated from an initial group of elements. Geometries consisting of intersecting planes, cylinders, cones, spheres, ellipsoids, hyperboloids or tori can be constructed with a suitable nodal point distribution. Intersection lines between these components are calculated automatically by the code. A user-defined quadratic line can be used to generate node placement. Any coordinate system can be used for specifying node positions, with a possibility of transfers between the systems and reflections about the axes of symmetry. Boundary conditions can be generated and plotted by the mesh preparation routines. Multiple sets of boundary conditions can be generated from a minimum of input data through the use of a pre-processing routine available for the code. For transient analysis, several thousand sets of input data may be generated with ease.

The library of finite elements available in the code associates with each element proper functional relationships for stress or thermal calculations. The number of elements formulated for each type of analysis is as follows:

- 40—static and dynamic mechanical loads
- 15—heat transfer
- 3—thermal–fluid interaction
- 3—thermal–electric interaction
- 3—fluid–solid interaction

Complete plastic, creep and swelling capabilities are provided for at least one element of each type listed above. Arbitrary stiffness, damping and mass elements are available and may be specified.

The choice of structural elements available in the code library include beams, plates, shells and solids. Pipes, elbows, axisymmetric and plane membranes are supplemented with elements representing springs, cables, dampers and sliding interfaces. In calculations involving thermal loading, heat transfer elements may be replaced by equivalent structural elements, for evaluation of thermal stresses.

The code structure permits the inclusion of elements with inelastic properties in static and nonlinear analyses. For structural elements in plane and solid configuration, orthotropic material properties may be specified. The same applies to elements used in heat transfer calculations. Curves of plastic stress–strain relationship can be input for up to five temperatures. Temperature-dependent material properties of elastic elements may be entered in the form of polynomial expressions with terms up to the fourth power of the temperature. If material properties are entered in tabular form, curve fitting and linear interpolation may be performed internally by the code. Temperature dependent convection film coefficients and emissivities for radiant heat transfer are allowed.

After the problem setup and checkout is complete, solution of the resulting matrix equations is carried out by means of the Guyan reduction, also known as the dynamic matrix condensation. It consists of an initial solution of a system with a reduced number of degrees of freedom, approximately characterizing the system solved. This master set of degrees of freedom for the mass, damping and stiffness matrices can be selected automatically by the code, or specified by the user. If desired, the solution can then be expanded to the full number of degrees of freedom of the entire problem, with the expenditure of additional computer time. A matrix condensation technique is available for evaluating the buckling, the mode-frequency and the reduced dynamic and harmonic frequency of response.

The results of analyses can be printed or plotted. Various arithmetic operations may be performed on the output in the post-processing stage, and extensive graphic capability is built into the code to facilitate display of three-dimensional geometries. Hidden line removal is available for displays of results. Stress–strain plots may be obtained from the results produced by the code.

This code is proprietary, and its use requires licensing arrangements with Swanson Analysis Systems, Inc., Johnson Rd., P. O. Box 65, Houston, PA 15342, tel. (412) 746-3304. Alternately, arrangements for the use of the code on a fee basis can be made through Babcock and Wilcox Computer Services Center, P. O. Box 1260, Lynchburg, VA 24505, tel. (804) 384-5111, ext. 3512.

REFERENCES

1. R. H. Gallagher, J. Padlog, and P. P. Bijlaard, "Stress Analysis of Heated Complex Shapes," *ARS J.*, 32, No. 5, 700–707 (May 1962).
2. G. A. Greenbaum and M. F. Rubinstein, "Creep Analysis of Axisymmetric Bodies Using Finite Elements," *Nucl. Eng. Des.*, 7, 379–397 (1968).
3. O. C. Zienkiewicz, *The Finite Element Method in Engineering Science*, McGraw-Hill, London (1971).
4. C. S. Desai and John F. Abel, *Introduction to the Finite Element Method: A Numerical Method for Engineering Analysis*, Van Nostrand Reinhold, New York (1972).
5. R. H. Gallagher, "Computational Methods in Nuclear Reactor Structural Design for High Temperature Applications: An Interpretative Report," ORNL-4756, Oak Ridge National Lab. (1973).
6. K. J. Bathe and E. L. Wilson, *Numerical Methods in Finite Element Analysis*, Prentice-Hall, Englewood Cliffs, NJ (1976).
7. Richard H. Gallagher, *Finite Element Analysis: Fundamentals*, Civil Engineering and Engineering Mechanics Series, Prentice-Hall, Englewood Cliffs, NJ (1975).
8. T. Belytschko, P. V. Marcal, and J. R. Osias, eds., "Finite Element Analysis of Transient Nonlinear Structural Behavior," AMD-Vol. 14, American Society of Mechanical Engineers, New York (1975).
9. R. C. Sampson, and R. E. Jagels, "Stress Analysis for the Design of Liquid Metal Piping in the Fast Flux Test Facility," ASME Paper No. 78-PVP-21 (1978).
10. M. Sherpherd and R. H. Gallagher, "Finite Element Grid Optimization," PVP-Vol. 38, American Society of Mechanical Engineers, New York (1979).
11. T. J. R. Hughes, ed., "Finite Element Methods for Convection Dominated Flows," AMD-Vol. 34, American Society of Mechanical Engineers, New York (1979).

12. Robert D. Cook, *Concepts and Applications of Finite Element Analysis,* John Wiley & Sons, New York (1981).
13. T. J. R. Hughes and D. Gartling, eds., "New Concepts in Finite Element Analysis," AMD-Vol. 44, American Society of Mechanical Engineers, New York (1981).
14. T. J. R. Hughes, ed., "Nonlinear Finite Element Analysis of Plates and Shells," AMD-Vol. 48, American Society of Mechanical Engineers, New York (1981).
15. "MARC-CDC, Nonlinear Finite Element Analysis Programs," Vols. 1–4, MARC Analysis Corp. and Control Data Corp., Minneapolis (1974).
16. "WECAN, Westinghouse Electric Computer Analysis, User's Manual," Advanced Systems Technology Division, Pittsburgh, PA (1976).
17. "WECAN: Users' Experiences," *Proc.* of the 3rd Users' Colloquium, Westinghouse Advanced Systems Technology Division, Pittsburgh, PA (1978).
18. Bathe, K. J., "ADINA—A Finite Element Program for Automatic Dynamic Incremental Nonlinear Analysis," AVL-82448-1, Dept. of Mech. Eng., MIT, Cambridge, MA (Sept. 1975; rev. Dec. 1978).
19. Deene J. Weidman, "NASA NASTRAN Structural Analysis Computer Program—New Content," NASA, Langley Research Center (1978).
20. "NASTRAN User's Guide—Level 15," CR-2504 (Apr. 1975).
21. "NASTRAN Users Colloquium," 9th NASA Structural Analysis Conf. Kennedy Space Center, FL, Oct. 22–23, 1980, Publ. NASA Scientific and Technical Information Branch, Washington, DC (1980).
22. Kohnke, P. C., "ANSYS—Engineering Analysis System, Theoretical Manual," Swanson Analysis Systems, Houston, PA (1977).
23. "ANSYS—Computer Program User's Manual," Version 671, Swanson Analysis Systems, Houston, PA (July 1979).

7.2 HEAT TRANSFER AND TRANSIENT ANALYSIS CODES

The development of nuclear reactors stimulated advances in the methodology and techniques for analysis of heat transfer, in conditions of laminar and turbulent flow of fluids in lattice channels or through fuel rod bundles. Both single channel and multichannel geometries were treated, with and without crossflow. Inherent in the nuclear technology was the desire to provide conditions sustaining maximum heat fluxes compatible with a high degree of safety. Although design concepts for heat transfer fluids initially covered the full range of possibilities, later there occurred the inevitable weeding out of less promising concepts. Starting with the use of compressed gases, through light and heavy water and slurries, to molten salts and liquid metals, the field narrowed to helium and carbon dioxide for gases, light and heavy water, and molten alkali metals. In reactors fueled with uranium ceramics, heat transfer specialists had to contend with the very low thermal conductivity of UO_2 and the consequent steep thermal gradients developed in the fuel rods.[1] Fuel pellet integrity, reduction in the width of the gas gap separating the fuel from its metallic cladding, and the maintenance of cladding integrity become problems of first importance.

The flow regimes considered for water included turbulent flow under conditions of local or bulk boiling. Problems of heat surface dryout, and the methods to prevent departure from nucleate boiling, came to occupy an important place in analytical and engineering considerations.[2] The work of Tong, Weisman, Jones, Fenech and others[3]–[6] contributed much to the advancement of the field. Almost from the inception of the nuclear industry, computer codes were written to facilitate calculations of heat transfer in geometries specific to nuclear reactors. Two such early codes were HECTIC and THINC,[7]–[8] quickly followed in succession by G-20, MANTA, SASS and COBRA.[9]–[12] Eventually a large family of computer codes was developed for predicting nuclear heat transfer. The code collection of the National Energy Software Center at Argonne National Laboratory contains 58 codes in this category. The code family developed from the early HECTIC program sparked considerable interest, and further work gained the sponsorship of ASME, leading to a series of topical symposia.[13] An example of more recently developed codes is the COMMIX code.[14]

Numerical analysis in thermo-hydraulics includes computer codes used for predicting normal reactor performance, and safety margins. The normal operating performance comprises the assessment of the nuclear, thermo-hydraulic and energy conversion aspects of reactor operation in the static regime. In the following section, characteristic features of better-known codes of that family will be discussed in more detail with reference to nuclear applications.

An experimental program supports the development of computer codes. In addition to experiments specifically designed to test accident conditions, regular power plant operating data are of great help. Such data, if properly tagged and annotated, can be used to check numerical simulations performed with accident analysis codes. These cross checks have been an integral part of the large verification program[15] initiated by the Electric Power Research Institute in connection with the development of the RETRAN code. The thoroughness of this verification program will make it a landmark in the field. In 1981 and 1982 two international conferences on RETRAN were held under EPRI sponsorship,[16],[17] to report the results of the verification. Another important code in the same category is the RELAP code. The features of both codes are discussed in more detail below.

The accident analysis codes discussed here, RETRAN and RELAP, depend heavily on the correct representation of reactor kinetics in their equations, combined with an accurate modeling of time-dependent heat transfer in the fuel bundles, as well as of the assembly inlet and outlet flow. The modeling of thermal and hydraulic inertia of the rest of the heat transfer and the energy conversion loop includes steam generators, turbines and condensers. In practice, the complexity of the equations involved means that very large computer codes are needed, and checking out their predictions becomes a major task. This, indeed, is the main motivation for the Loss of Fluid Test (LOFT) series of experiments currently in progress at the Idaho Reactor Testing Station.

Several other large codes for system simulation under transient or accidental conditions are under development: the RAMONA[(18)] code for boiling water reactor systems: the CONTEMPT code[(19)] used with modifications to predict pressure response in BWR containment enclosures of the suppression type, following the loss-of-coolant accident; the TRAC code[(20)] developed in 1969 by T. F. Peterson of Gulf General Atomics for general, two-dimensional, heat transfer analysis in reactors. This code has undergone many extensions, particularly at the Los Alamos Laboratory,[(21),(22)] and is now often used for loss-of-coolant analyses.

Recently major computer manufacturers started providing software for thermo-hydraulic calculations. One of the best known is the CYBERNET Service of the Control Data Corporation.

COBRA Code

The COBRA Code[(23)] in its first version was developed by D. S. Rowe of Battelle Northwest Laboratories at Hanford. Since that time numerous changes and improvements have been made in it at the author's laboratory[(24)–(25)], and at MIT.[(26)] The latest version was developed by the Argonne National Laboratory.[(27)] Although the code was intended as a general purpose steady state heat transfer code, its later versions are tailored more toward the specifics of fast reactors, making it one of the very popular analytical tools for steady state and transient heat transfer analyses in fuel rod bundles for Liquid Metal Breeder Reactors. The latest versions of the code have input, output and internal calculations carried out in SI units.

The COBRA code performs thermal-hydraulic analysis of a fuel bundle for a nuclear reactor loaded with fuel pins held in a grid. The bundle cross section is divided into coolant subchannels, fuel pin sectors and duct wall sectors for the purpose of analysis. Convection of heat in the direction of coolant mass flow is calculated, as well as heat flow between pin sectors and adjoining coolant subchannels, in addition to heat and mass transfers between coolant channels themselves. Asymmetries of pin-to-pin and pin internal heat release rates are accounted for in a sophisticated thermal model of the assembly. Heat losses to the outer duct wall are also considered. The method of solution is based on the finite difference equations, solved separately for:

- Inner subchannel cross flow
- Inner subchannel heat transfer
- Heat transfer within each pin sector
- Heat transfer within each duct wall sector

Computations of neighboring regions are related by consistent boundary conditions for heat fluxes. Equations are solved for one axial mode at a time, starting at the bottom of the fuel bundle. Coolant flow and temperature field at the top of each axial node serve as a boundary condition for the node directly above. Iteration is used to converge both the steady state and the transient solutions. The internal logic of the latest version of the code has been streamlined to improve the speed of matrix inversions.

The melting transition of the fuel is included in the code model, as well as a sophisticated fuel cladding gap conductance formulation. Through the use of the computed energy balance, the user can judge the adequacy of the chosen convergence criteria for flow and temperature.

Through the dynamic dimensioning of arrays used within the code, considerable latitude is provided for the analyst to adjust the problem to the capability of the available computer memory size. Free-formatted input is allowed, restarts of long runs of the same problem are possible, and linkage of COBRA to other programs is available through user routines. It is, for instance, possible to provide the distribution of coolant temperatures as output, to be used as input to a code computing fuel behavior.

The latest version of the code reduces the computer memory storage requirements by the use of overlays, elimination of unnecessary arrays, and selective choice of variables requiring double-precision arithmetics. Analysis of a 37-pin fuel rod bundle with the COBRA-PI version of the code on an IBM 370/195 machine requires 238k words of random access memory. A CDC-7600 version of the code is also available.

REPP Code

The REPP Code[28] was written by R. M. Hiatt and C. Bromley of Pacific Northwest Laboratory at Hanford, to calculate the heat transfer performance of the core lattice of a water-cooled reactor. The code evaluates fuel temperatures and the critical heat flux margins required in thermal design. The code model does not fully treat the effect of restricted flow in the corners of a fuel assembly, or the influence of a cold wall on the burnout heat flux.

For a given specific margin between the operating and the burnout heat flux, the code determines the number of fuel pins required for the core to achieve a specified power level. The code will determine the diameter of the fuel pin required to operate within specified center line temperature limits at a given core power level, and will evaluate the effects of reorganization of the fuel crystalline structure on the fuel temperature distribution. It will calculate the pressure drop and the properties of the coolant in single- and two-phase flow at both average and hot channel fuel pin conditions, and it will also calculate pressure drop across several types of fuel pin spacer grids using the A. N. deStordeur technique. The choice of spacers is: a spiral wire wrap, a honeycomb spacer or lenticular wire spacers.

The reactor core is represented in the REPP model as a right circular cylinder made up of fuel pins arranged in a square or triangular lattice. The vertical flow channels are divided into no fewer than 10 axial mesh nodes, and no more than 100. Within each fuel rod 21 radial nodes are specified for the evaluation of the temperature profile between the fuel rod center line and its surface. A provision for the temperature-dependent variation of thermal conductivity of fuel is incorporated in the code model. Radial flux depression in the fuel may be specified in the input independently of the nodal mesh. For oxide fuels operating at sufficiently high heat fluxes, the formation of the central void in the fuel pin may be specified, and the resulting void diameter will be calculated.

During the calculation, an energy and momentum balance is made at each nodal point, and thermal and hydraulic conditions are calculated for the nominal and the hot channels. A subroutine incorporating the Keenan and Keyes steam tables is provided for the calculation of coolant properties at each node. W-2 and W-3 burnout correlations are built into the code. Integrations required by the code are performed by Simpson's rule. The code input requires specification of the normalized axial power profile and the total reactor power. Since the axial power is specified independently of the nodal mesh, a Lagrangian interpolation routine is provided to facilitate the transition.

The code is written in FORTRAN V, and is mechanized on the UNIVAC-1108 machine with CSCX operating system. For operation it requires 51k words of addressable magnetic core storage space. An iterative solution can be obtained in 30 seconds.

RETRAN Code

The RETRAN Code[29] is the end result of a development effort initiated in 1975 by the Electric Power Research Institute. It is a one-dimensional homogeneous equilibrium thermo-hydraulic code developed from the RELAP-4 series of codes. Originally released in 1978, it is now available in version two.[30] RETRAN is a general purpose code for steady state and transient analysis of boiling water and pressurized water reactors. The code was primarily developed as a tool to be used by utilities in evaluating and improving design and operation of reactor systems. It is intended for evaluation of safety considerations and support of licensing applications.

The code produces best-estimate predictions of a thermal-hydraulic transient in a nuclear steam supply system. Both normal operational conditions and hypothetical accident conditions can be analyzed. The latter may include loss-of-fluid accidents following a small or a large break in piping. Analysis of anticipated transients without control rod scram is one of its capabilities. The code can also be used for analyzing steady state and transient responses of any thermal-hydraulic system, not necessarily a nuclear reactor. The code includes a one-dimensional kinetics formulation to simulate fuel rod axial power feedback, and offers several optional feedback and control system functions. The water property data used by the code are in the form of polynomial fits, as a function of pressure and enthalpy, of the values contained in the 1967 ASME steam tables.

Specialized spatial finite differencing methods are used in RETRAN to discretize the three field equations: the one-dimensional integral homogeneous equation of continuity, the momentum conservation equation and the energy balance equation. For situations that involve convection of vapor and liquid coolant at unequal velocities, an optional field equation mechanizing the dynamic slip model is provided. The mixture mass balance equation is unchanged by slip effects, but the mixture momentum and the energy equations contain terms accounting for the unequal phase velocity effects. The field equations of the code revert to the homogeneous equilibrium model in situations when the dynamic slip model is not used. Separate solution techniques are used to solve thermal hydraulics, heat conduction, point kinetics, one-dimensional kinetics and control system equations. These equations are solved using either a Newton-Ralphson predictor-corrector integration technique, or an improved version of the Porsching method.[31]

The formulation of the code is perfectly general, giving the user an opportunity to assemble the basic building blocks specifying fluid control volumes, flow paths and junctions, heat conductors, cetrifugal pumps, valves and control systems, as well as nuclear or electrical power sources, until a complete description of the system is obtained. Addition or deletion of the building block components is possible, because the mode or component numbering assignment is random. Figure 7 shows a typical RETRAN model, as applied to the TMI plant.

The actual code consists of four modules:

- RETRAN, best-estimate transient thermal-hydraulic analysis code incorporating the steady state option, which computes steady state initial values of all parameters of interest.
- RESTRT program for continuing the computation after halts to examine partial results.
- REEDIT program used to generate printed edits of data stored on tape by the first two modules.
- PLOTTER program to generate Calcomp-type plots from output data stored on tape.

The code is available in versions mechanized on the CDC-6600 computer, or the IBM-360 and 370 systems. Running time is problem-dependent, and may vary from a few minutes to a few hours. The code is proprietary, and EPRI licenses are required for use both within and outside of the United States.

RELAP Code

The RELAP-5 Code is the fifth version of a code developed in the late 1960s[32] by K. V. Moore and W. H. Rettig of the USAEC Idaho Laboratory. It incorporates a good part of the core blowdown logic of the earlier codes FLASH and RELAPSE. The development of the later versions of the code[33]–[35] was done under the sponsorship of the Nuclear Regulatory Commission. It is now one of the principal codes used for the analysis of transients induced in water-cooled reactors by loss-of-coolant accidents, pump failures or power excursions.

The latest version of the RELAP program calculates coolant flow, pressure, mass inventory and quality. Thermal conditions of heat transfer such as surface temperatures and their profiles, energy distributions and heat fluxes in the power generating and dissipating elements of the system are determined. Reactor power, decay heat and reactivity of the core are also calculated. The formulation of the code is general enough so that not only can transients be analyzed in pressurized water and boiling water reactors, but in experimental thermal-hydraulic systems as well. For the analysis of PWR core reflood phenomena, a special model of RELAP, RELAP/MOD6, was created.

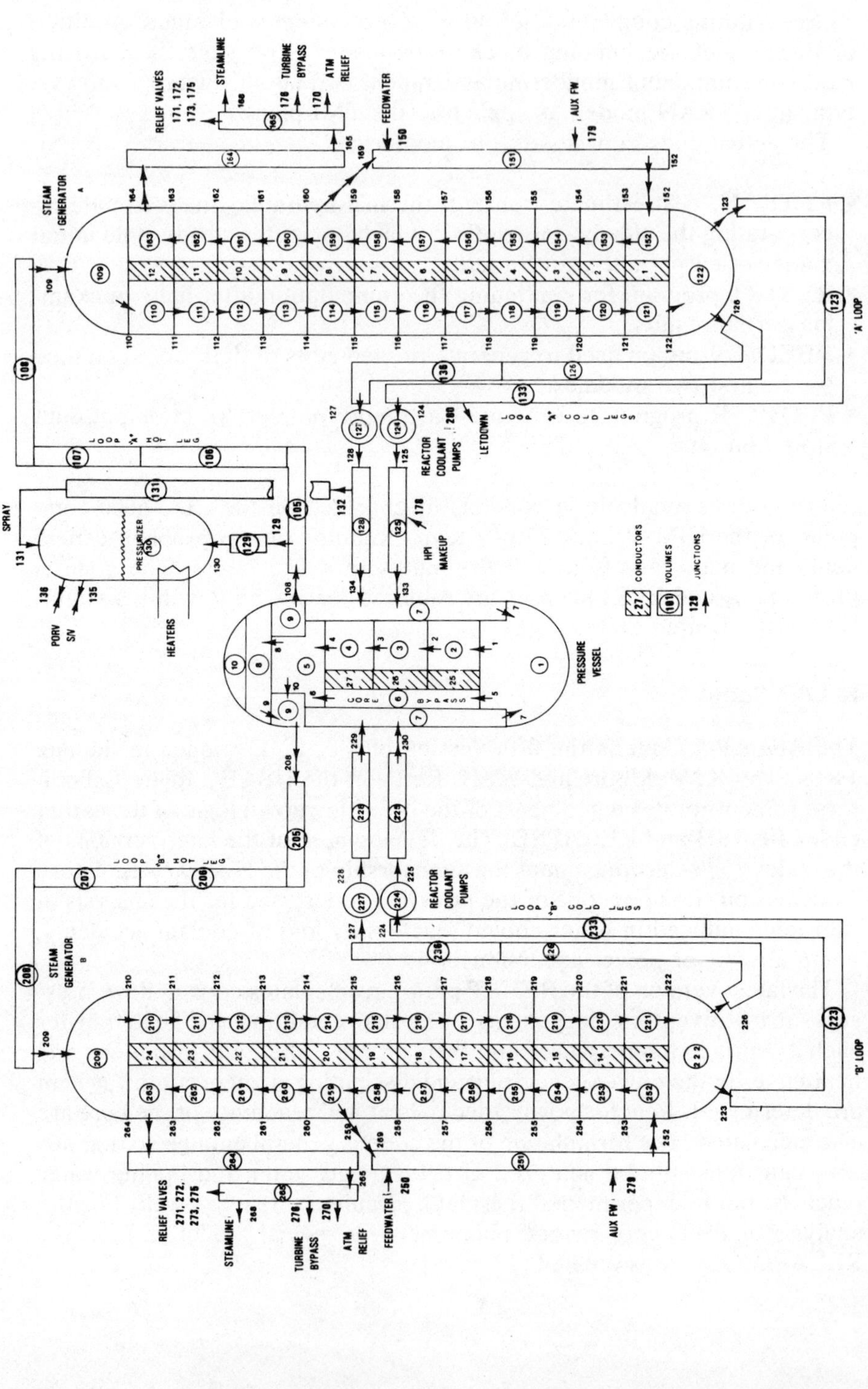

Figure 7. Two-loop RETRAN model for the Three Mile Island Unit 2. (Reproduced courtesy GPU Nuclear Corporation)

In solving these problems, RELAP calculates an integral form of fluid conservation and state equations for each user-defined element of the system. A fully implicit integration technique is used. Steam tables are built into the program and contain derivatives of state properties. Heat conduction between fluid volumes is allowed, and momentum flux terms are included in the hydraulic equations. The pump model incorporated in the code has explicit inertia and friction properties. A separate hot channel calculation is done for the core heat transfer, and cross flow in the core is allowed.

Version four of the program contains the following operational blocks:

- RELAP4: standard model with minimum controls
- RELAP4-EM: evaluation model of NRC
- RELAP4-FLOOD: model with reflooding rates
- RELAP4-CONTAINMENT: model with containment representation

The post-processing code PLOTR4M produces hard-copy printed output as well as CalComp and microfilm plots of important variables versus time. The pre-processing code PRELOAD reduces core storage requirements of the main code, while STH20G generates tables of water properties in proper format for the IBM-360 version of the code.

The code is mechanized on the CDC-7600, IBM-360 and IBM-370. The random access magnetic core memory size required is $160k_8$ for Small Core Memory and $200k_8$ for Large Core Memory of the CDC-7600 machine. For the IBM-360, 850k bytes are required. The typical running time for a PWR model with 30 to 40 fluid volumes is 15 to 60 minutes on the CDC-7600 and from 2 to 8 hours on the IBM-360.

REFERENCES

1. Jacek Jedruch, "The Effect of Radial Temperature Gradient on the Formation of Axial Void in UO_2 Fuel Elements," Int. J. Heat Mass Transfer, 11, 1155–1165 (1968).
2. L. S. Tong, *Boiling Crisis and Critical Heat Flux,* U.S. Atomic Energy Commission (1972).
3. L.S. Tong and J. Weisman, *Thermal Analysis of Pressurized Water Reactors,* American Nuclear Society, La Grange Park, IL (1970).
4. O. C. Jones, Jr., *Nuclear Reactor Safety Heat Transfer,* Hemisphere/McGraw-Hill, New York (1981).
5. P. Bradshaw et al., *Engineering Calculation Method for Turbulent Flow,* Academic Press, New York (1981).
6. Henri Fenech, ed., *Heat Transfer and Fluid Flow in Nuclear Systems,* Pergamon Press, New York (1982).
7. N. Kattchee and W. C. Reynolds, "HECTIC II, An IBM 7090 Fortran Computer Program for Heat Transfer Analysis of Gas or Liquid Coolant Reactor Passages," IDO-28595, Aerojet General Nucleonics (Dec. 1962).

8. W. Zernick, H. B. Currin, E. Elyash, and G. Previti, "THINC, A Thermal Hydrodynamic Interaction Code for Semi-open or Closed Channel Core," WCAP-3704, Westinghouse Electric Co. (Feb. 1962).
9. I. E. Oldaker, "Sheath Temperatures, A G-20 Computer Program for Fuel Bundles," AECL-2269, Chalk River Labs. (May 1965).
10. S. F. Armour and D. L. Smith, "MANTA—Mixing Analyzed Nodal Thermal-Hydraulic Analyses," GEAP-4805, General Electric Co. (Feb. 1965).
11. C. C. St. Pierre, "SASS Code I—Subchannel Analysis for the Steady State," AECL-APPE-41, Atomic Energy of Canada Ltd. (Sept. 1966).
12. D. S. Rowe, "COBRA—Computer Program for Coolant Boiling in Rod Arrays," BNWL-371, Pt. 1, Battelle Northwest Labs. (Mar. 1967).
13. "Heat Transfer in Rod Bundles," *Proc.* New York Symp. Heat Transf. Div., American Society of Mechanical Engineers, New York (Dec. 3, 1968).
14. W. T. Sha, H. M. Domanus, et al., "COMMIX-1: A Three-Dimensional Transient Single-Phase Component Computer Program for Thermal Hydraulic Analysis," NUREG/CR-0785, ANL-77-96, Argonne National Lab. (Sept. 1978).
15. "Comparison of Experimental Results with RETRAN Predictions for LOFT Test L2-3 (NRC Standard Problem 10)," EPRI NP-1352, Intermountain Technologies, Inc. (1980).
16. "Proceedings of the First International RETRAN Conference," EPRI WS-80-150, Electric Power Research Institute (Apr. 1981).
17. "Proceedings of the Second International RETRAN Conference," EPRI NP-2494-SR, Electric Power Research Institute, (July 1982).
18. D. J. Diamond, C. J. Ruger, D. I. Garber, and S. V. Lekach, "RAMONA-3B, A Code for BWR Plant Transients," *Proc.* Conf. Simulation Methods for Nuclear Power Systems, WS-81-212, Electric Power Research Institute (1981).
19. L. L. Eheat et al., "CONTEMPT-LT, A Computer Program for Predicting Pressure–Temperature Response to a Loss-of-Coolant Accident," ANCR-1219, Aerojet Nuclear Co. (1975).
20. T. F. Peterson, "TRAC2D—A General Purpose Two-dimensional Heat Transfer Computer Code—User's Manual," GA-8868, Gulf General Atomics (1969).
21. John C. Vigil and Richard J. Pryor, "Development and Assessment of the Transient Reactor Analysis Code (TRAC), *Nuclear Safety,* 21, 171–183, (Mar.–Apr. 1980).
22. Safety Code Development Group, "TRAC-PD2: An Advanced Best-Estimate Computer Program for Pressurized Water Reactor Loss-of-Coolant Analysis," LA-8709-MS, NUREG/CR-2054, Los Alamos National Lab. (Apr. 1981).
23. Rowe, D. S., "Cross Flow Mixing Between Parallel Flow Channels During Boiling," (Part I) "COBRA—Computer Program for Coolant Boiling in Rod Arrays," BNWL-371, Pt. 1, Pacific Northwest Lab., Hanford, WA (Mar. 1967).
24. T. L. George, K. L. Baseshore, C. L. Wheeler, W. A. Prather, and R. E. Masterson, "COBRA-WC: A Version of COBRA for Single Phase Multiassembly Thermal-Hydraulic Transient Analysis," PNL-3259, Pacific Northwest Lab. (July 1980).
25. E. P. Coomes, E. U. Khan, T. L. George, R. L. Cheetham, and D. R. Rector, "User Manual: COBRA-WC: A Version of COBRA for Single-Phase, Single and Multiassembly Thermal-Hydraulic Transient Analysis," PNL-4303, Pacific Northwest Lab. (May 1982).
26. J. W. Jackson and N. E. Todreas, "COBRA-IIIc/MIT-2: A Digital Computer Program for Steady State and Transient Thermal-Hydraulic Analysis of Rod Bundle Nuclear Fuel Elements," MIT-EL 81-018, Massachusetts Institute of Technology (June 1981).
27. T. H. Bauer, and P. H. Froehle, "COBRA-PI: An Extension of the COBRA-3M Code Dynamically Dimensioned to Accept Pin Bundles of Any Size," *Proc.* of ANS Meeting

on Advances in Reactor Computations, Vol. I, 211–213, Salt Lake City (Mar. 28–31, 1983).
28. R. M. Hiatt and C. Bromley, "REPP: A Thermal-Hydraulic Design Code for Water-Cooled Reactors," BNWL-1013, Pacific Northwest Lab. (Mar. 1969).
29. K. V. Moore et al., "RETRAN-01: A Program for One-Dimensional Transient Thermal-Hydraulic Analysis of Complex Fluid Flow Systems," Vol. 1: Equations and Numerics, Vol. 2: Programmer's Manual, Vol. 3: User's Manual, Vol. 4: Applications, EPRI CCM-5, Electric Power Research Institute (May 1978).
30. J. H. McFadden et al., "RETRAN-02, A Program for Transient Thermal-Hydraulic Analysis of Complex Fluid Flow Systems," Vols. 1–3, NP-1850-CCM, Electric Power Research Institute (May 1981).
31. M. C. Paulsen, J. H. McFadden, and C. E. Peterson, "Review of the Solution Approaches, Methods, and Recent Results of the RETRAN-02 Systems Simulation Code," *Proc.* of ANS Topical Meeting on Advances in Reactor Computations, Vol. 2, 854–875, Salt Lake City (Mar. 28–31, 1983).
32. K. V. Moore and W. H. Rettig, "RELAP2—A Digital Program for Reactor Blowdown and Power Excursion Analysis," IDO-17263, USAEC, Idaho Operations Office (Mar. 1968).
33. W. H. Rettig et al., "RELAP-3, A Computer Program for Reactor Blowdown Analysis," IN-1321 (June 1970).
34. "RELAP4/MOD5: A Computer Program for Transient Thermal-Hydraulic Analysis of Nuclear Reactors and Related Systems," Vols. 1–3, ANCR/NUREG/1335 (Sept. 1976).
35. V. H. Ransom et al., "RELAP5/MOD1 Code Manual, Volume 1: System Models and Numerical Methods, and Volume 2: Users Guide and Input Requirements," NUREG/CR-1826 (EGG-2070), EG&G Idaho (Apr. 1982).

7.3 FAST REACTOR SAFETY ANALYSIS CODES

Some of the computer codes described in the preceding section for transient analysis of water reactors, such as RETRAN, are applicable to the analysis of early stages of flow accidents. For fast reactors, however, a special analytical tool had to be developed.

For LMFBR safety analyses specially adapted codes are needed to represent adequately the differences in the physics of fast breeders compared to thermal reactors. These differences appear in the magnitude, and possibly sign, of the void and temperature coefficients of reactivity, and the reduced neutron lifetime parameters. Their effects are particularly pronounced in the fast reactor response to the initiating reactivity perturbations. The response of liquid metal hydraulic systems is also markedly different. A transient power rise and any reduction in the density of the liquid metal coolant in the core affect the magnitude of neutron leakage from the core, and have a feedback effect on the duration and the severity of the power transient.

To handle the differences in core physics, a respectable group of computer codes has been developed for accident analysis of LMFBRs. Representative of this group are such computer codes as SSC-L,[(1),(2)] SAS[(3)] and

SIMMER.[4] Computer codes in this category share certain general features. Among them is a fairly detailed model of the fuel pin, required for computation of a detailed temperature distribution. This aspect of computation is especially important for fast reactors because of their higher fissile loading of fuel, and the subsequent reduction of the negative component of the Doppler effect. When power density is sufficiently high to induce melting of fuel cladding and the disintegration of fuel, any additional energy releases produced by interaction between the molten fuel and other components of the core need to be represented in the fuel model as well.

SAS Code

The SAS_{1A} Code[3] was developed in 1969 by A. K. Agrawal and coworkers at Argonne National Laboratory for analysis of fast reactor power and flow transients. The code also has the capability of estimating the consequences of a weak explosion in reactor accidents. The code exists in several versions with successively increasing complexity of modeling of geometry and the physical processes taking place in reactor transients. An interface to the SIMMER-II code is available,[4] and can be used for analyzing the consequences of accidents computed with SAS.

The neutron kinetic model built into the code is based on six delayed neutron groups. Either the variation of core reactivity or core power can be input as a function of time.

The heat transfer model of the code has 20 axial nodes placed along the fuel pin, including the upper and lower blanket regions. The fuel pin is represented by 15 radial temperature nodes, while the core is subdivided into 60 radial regions.

The motion of fuel cladding in an accident sequence is described in the Lagrangian system in SAS3D. A German version of the code,[5] developed at Karlsruhe, contains a reformulation of the model of cladding motion in the Eulerian system, and so is able to register even small displacements of the cladding.

The code is organized in the main calling program and four modules as follows:

- Input–output with diagnostics
- Steady state initialization of analysis
- Transient analysis module
- Weak explosion module to simulate core disassembly

The output of the code can be used as input to the SIMMER-II code for a more detailed analysis of the disrupted core and further consequences of accidents.

SAS is mechanized on the IBM-360/75 machine and requires 410k bytes of random access memory, when loaded into the machine without overlays, and 300k bytes with overlay loading. The running time for a typical problem is five minutes.

SIMMER Code

The SIMMER Code[(6)] was developed at the Los Alamos National Laboratory for the analysis of consequences of core disruption of a fast breeder reactor. In the code the space- and time-dependent neutronics calculation is coupled with multiphase, multicomponent Eulerian fluid dynamics. It is intended for the calculation of two-dimensional (r,z) motion of core materials during a disruptive LMFBR accident. The code model handles turbulent flow, and accounts for radiative energy transport.

A quasi-static method is used for calculating code neutronics, with a two-dimensional, finite-difference technique applied in either the diffusion or the transport theory option. Alternatively, a point kinetics calculation can be specified as input. Isotropic scattering is assumed, and self-shielded and temperature dependent cross sections can be specified by region. No account is taken of the transport of delayed neutron precursors.

In the calculation of fluid dynamics a coupled set of multicomponent, multifield equations is solved for the conservation of mass, momentum and energy. Nine density and five energy components are used in the structure field. In the liquid field eight density and six energy components are used, while in the vapor field six density and five material components are used. Analytic and tabular formulations of the equation of state are available as options. A combination of explicit and implicit time differencing techniques is used together with donor-cell spatial differencing. To calculate the interfacial phenomena between components the dispersed droplet flow regime is used. Intact and partially disrupted subassemblies are represented with a simple fuel pin model without mechanical deformation.

The source deck of the code consists of approximately 25,000 lines, and the code is arranged into 32 overlay blocks.

In the execution of the program a selection of various option blocks is possible through an update routine. Output of the program is available on paper, microfilm, microfiche or a movie film.

Although the code was programmed for the CDC-7600 machine, it has been run on IBM and CRAY machines. On CDC a minimum of 55k Small Core Memory and between 50k and 280k Large Core Memory are needed. Running time is greatly dependent on the options used.

MARCH Code

The MARCH Code[7] was developed by R. O. Wooton and H. I. Avci to model in a consistent manner the physical processes of core meltdown. It is a versatile code modeling a variety of processes. It is used in probabilistic risk analyses that require an investigation of a large number of accidental sequences. For this reason it is important that the code be economical to run, with the implicit assumption that approximate solutions to many complex phenomena will be obtained.

REFERENCES

1. A. K. Agrawal, "An Advanced Thermohydraulic Simulation Code for Transients in LMFBRs (SSC-L Code)," BNL-NUREG-50773, Brookhaven National Lab. (1978).
2. "Compendium of Computer Codes for the Safety Analysis of Fast Breeder Reactors," DOE/ET-0009, 1st edit., Department of Energy, Washington, DC (Oct. 1977).
3. A. K. Agrawal et al., "SAS1A, A Computer Code for the Analysis of Fast Reactor Power and Flow Transients," ANL-7607, Argonne National Lab. (Aug. 1969).
4. R. G. Steinke and P. J. Hodson, "An Automated Interface from SAS3D to SIMMER-II," NUREG/CR-1516 (Oct.–Dec. 1979).
5. G. Angerer, G. Arnecke, and A. Polch, "SAS3DC: A Computer Program to Describe Accidents in LMFBRs," KFK-3101, Kernforschungszentrum Karlsruhe (Feb. 1981).
6. L. L. Smith, "SIMMER-II: A Computer Program for LMFBR Disrupted Core Analysis," NUREG/CR-0453, LA-7515-M, Los Alamos National Lab. (Oct. 1978).
7. R. O. Wooton and H. I. Avci, "MARCH (Meltdown Accident Response Characteristics) Code Description and User's Manual," NUREG/CR-1711, BMI-2064 (Oct. 1980).

7.4 PROBABILISTIC RISK ASSESSMENT

In the field of reactor safety analysis, the probabilistic risk assessment (PRA) is now in the process of vigorous development. Using estimates of expected failure frequencies for individual reactor components, combined through a fault and event tree with probabilities of failure of other components, the method obtains numerical estimates of the probability of failure of the whole system in a given operating interval. The method received initial acclaim for its application to the reactor safety study conducted on behalf of NRC by Norman Rasmussen of MIT.[1]

In the approach used by Rasmussen, and later elaborated by others, the combined effect of various equipment failures is tracked in a computer program to the various possible outcomes. This last part of the probabilistic risk analysis is often termed consequence analysis. This method requires that a sufficiently large bank of data on the rate and frequency of mechanical failures be available to be sampled along each branch of the fault tree. For its successful application, the method requires the ability to manipulate data from failure data sources, to select suitable probability distributions to represent that data, and finally to convert the data into expected availability of systems of linked components. The preparation of data bases containing failure data, reported from various reactor installations, and the status of their holdings is discussed in another chapter. An excellent introduction and summary of this method was presented recently by McCormick.[(2)]

To provide a measure of systematization in the field of PRA, the Nuclear Regulatory Commission prepared a procedures guide[(3)] which recognizes three levels of application of PRA:

1. Only failure event sequences are estimated without calculation of consequences of failure. The objective at this level is to investigate the reliability of vital plant components.
2. Analysis of in-plant consequences of accident sequences is carried out to define the probability of radioactivity releases to the containment.
3. Ex-plant consequences of failure chains are also considered, with particular reference to the health and safety of the public.

Also available from NRC is a "Fault Tree Handbook."[(4)] The handbook summarizes the theory on which this method is based, and discusses the computer codes used to enable its practical implementation.

The efforts at computerizing the fault tree analysis produced a better definition of methodology[(5)–(7)] and the development of a number of computer codes.

Taking advantage of the fact that failure processes are basically random in nature, but with a frequency distribution unknown a priori, Monte Carlo computer codes were written to allow their simulation and subsequent comparison with experiments. This approach is well illustrated by the EXPECT Code,[(8)] developed in connection with the design and operation of devices for controlled thermonuclear research.

The analytical approach to the estimation of lifetimes of systems consisting of many replicate components led to the machine application of Weibull[(9)] statistics in this area. Various experimental programs were carried out to obtain the required data base, particularly in projects connected with the application of nuclear reactors to the propulsion of space vehicles.[(10)]

The first computerized fault tree evaluation codes PREP and KITT[11] were developed at the Idaho National Laboratory and became available in 1970. The first is a minimal path set generator, the second mechanizes the time dependent kinetic tree theory. Both are written for IBM-360 machine in FORTRAN IV. As is often true of early codes, they require large amounts of machine time to evaluate complex fault trees.

The next generation of PRA codes became available in mid 1970s and included SAMPLE and MOCARS. The first is a Monte Carlo code in which the component failure rates can be represented as normal, log-normal or log-uniform distributions. This code was used in the Rasmussen study leading to the WASH-1400 report. The code estimates the mean and standard deviation of the failure distribution as well as the histograms of results. MOCARS code[12] uses similar methodology but extends the choice of failure sampling distributions to include binormal, gamma, beta, Poisson and Weibull distributions as well.

Among other codes developed for PRA there are MOCUS, ALLCUTS, TREEL and MICSUP.[13]–[15] Sandia National Laboratory contributed the code SETS,[16],[17] and the Electric Power Research Institute supported the development of the WAM-BAM and the WAMCUT codes.[18],[19] A comparatively recent arrival is the second version of the COMCAN code.[20] Features of three codes representative of this group of codes are discussed below.

SETS Code

The SETS code,[16],[17] or the Set Transformation System, was developed by R. B. Warrell of the Sandia National Laboratory to facilitate the operation with the Boolean equations representing the fault trees. The code is formulated in a general manner, with a large degree of flexibility available to the user in the dynamic manipulation of large fault trees. An automatic feature of the code enables the identification of independent subtrees and makes it possible to process large event trees in stages without modifying the description of the original fault tree. Using any valid Boolean expression the user can define complemented events, and exclusive or special gates. Set equations involving the operations of union, intersection and complement are acceptable for input. Prime implicants of any intermediate gate can also be found.

Other convenient features of the code include the free field input format, the possibility of entering replicated subtrees, and the ability of saving the factored equations as a file, for input in subsequent calculations. From these equations path sets of any order can be reproduced for display or enumeration. If basic event probabilities have been included in the input, the code will rank and print in descending order of probability the minimal path sets.

The generic cause susceptibilities for each basic event can be input into the Boolean equations for the top or intermediate gates of the fault tree, using a variable transformation. The code can then be used to analyze common cause failure situations. With some simple manipulations the code will produce a display of cut sets which are common cause candidates.(16)

The SETS code is written in extended FORTRAN for CDC-6600. The running time of the code is strongly dependent on the number of expansion, simplification and factorization operations required for the given problem.

WAM-BAM Code

The WAM-BAM Code(18),(19) was developed under the sponsorship of EPRI by F. L. Leverenz and associates of Science Applications, Inc. It consists of four codes, the first two of which, WAM and WAMTAP, are input preprocessors, BAM (Boolean Arithmetic Model) is the main evaluation code, and the WAM-CUT is the postprocessor.

This code quantifies the system model in a single step. It allows the recognition of complement events and modelling of dependencies. It has the advantage of reducing the computer storage requirements by eliminating paths of low probability at an intermediate stage of processing. The total of discarded probabilities is, however, recorded.

The WAM preprocessor generates numeric input for the BAM module from the description of fault tree and the event probabilities. The probabilities for all components must be input, regardless of the significance of their contribution to system failure. This is one of the disadvantages of the code, since a change in probabilities may require a complete rerun of the code. Output from WAM preprocessor can be saved for later modification by WAMTAP for sensitivity studies. WAMTAP allows changing the probability of malfunction of single component, or a group of components, to include common cause contributions.

The BAM module computes the point probability of top event from a truth table in which each line represents a product term disjoint from all the other terms. The operations are modeled as gates on a fault tree, with eight possible logical combinations of two events and their complements included as allowable gates.

WAM-CUT is the module computing the mean and variance of the probability of any gate, as well as the minimal cut sets. The code is written for CDC-6600 in FORTRAN.

EXPECT Code

The EXPECT code[8] gives the designer a tool with which to predict failure rates of components, with special emphasis on the prediction of expected time to the next failure. The code permits the calculation of this quantity for a number of theoretical distributions, in order to allow a selection of the best distribution to fit any available experimental results. The distributions considered are: exponential, Weibull, Erlang, normal (actually, truncated normal), gamma, uniform.

The exponential and Erlang distributions are actually special cases of the gamma distribution.

REFERENCES

1. Norman Rasmussen, ed., "Reactor Safety Study—An Assessment of Accident Risks in U.S. Commercial Nuclear Power Plants," WASH-1400, NUREG-75/014, Nuclear Regulatory Commission (Oct. 1975).
2. N. J. McCormick, *Reliability and Risk Analysis,* Academic Press, New York (1981).
3. "PRA Procedures Guide," NUREG/CR-2300, Nuclear Regulatory Commission, Washington, DC (1982).
4. W. E. Vesely, F. F. Goldberg, N. H. Roberts, and D. F. Haasl, "Fault Tree Handbook," NUREG-0492, Nuclear Regulatory Commission (Jan. 1981).
5. J. B. Fussell, "Synthetic Tree Model—A Formal Methodology for Fault Tree Construction," ANCR-1098, Aerojet Nuclear Company (Mar. 1973).
6. G. R. Burdick and J. Fussell, "On the Adaptation of Cause and Consequence Analysis to U.S. Nuclear Power Systems Reliability and Risk Assessment," ANCR-1273, Report V, Aerojet Nuclear Company (Apr. 1976).
7. T. Smith et al., "A Risk-Based Fault Tree Analysis Method for Identification, Preliminary Evaluation, and Screening of Potential Accidental Release Sequences in Nuclear Fuel Cycle Operations," BNWL-1959, Battelle Northwest Labs. (Jan. 1976).
8. G. P. Boicourt, "EXPECT—A Monte Carlo Program to Predict the Expected-Time-to-Next-Failure in Controlled Thermonuclear Research Systems," LA-5980, Los Alamos Scientific Lab. (1975).
9. W. W. Weibull, "A Statistical Distribution Function of Wide Applicability," *J. Appl. Mech.,* 18, 293–297 (1951).
10. G. J. DeSalvio, "Theory and Structural Design Applications of Weibull Statistics," WANL-TME-2688, Westinghouse Astronuclear Lab., Pittsburgh PA (1970).
11. W. E. Vesely et al., "PREP and KITT: Computer Codes for the Automatic Evaluation of a Fault Tree," IN-1349, Idaho Nuclear Corp. (1970).
12. S. D. Matthews, "MOCARS: A Monte Carlo Simulation Code for Determining the Distribution and Simulation Limits," TREE-1138, (July 1977).
13. J. B. Fussell, E. B. Henry, and N. H. Marshall, "MOCUS—A Computer Program to Obtain Minimal Sets from Fault Trees," ANCR-1156, Aerojet Nuclear Company (Mar. 1974).
14. W. Van Sylke and D. Griffing, "ALLCUTS, A Fast Comprehensive Fault Tree Code," ARH-ST-112, Atlantic Richfield Hanford Co. (July 1975)
15. P. K. Pande et al., "Computerized Fault Tree Analysis: TREEL and MICSUP," ORC-75-3, Operation Research Center, University of California, Berkeley, CA (1975).

16. R. B. Warrell, "Common Cause Analysis Using SETS," SAND-77-1832, Sandia National Lab., Albuquerque, NM (Dec. 1977).
17. R. B. Warrell and D. W. Stack, "A SETS User's Manual for the Fault Tree Analyst," SAND-77-2051, Sandia National Lab. (1978).
18. F. L. Leverenz and H. Kirch. "User's Guide for the WAM-BAM Computer code," EPRI 217-2-5, Science Applications, Inc. (Jan. 1976).
19. R. C. Erdmann, F. L. Leverenz, and H. Kirch, "WAMCUT, A Computer Code for Fault Tree Evaluation," EPRI-NP-803, Science Applications, Inc. (June 1978).
20. D. M. Rasmuson et al., "COMCAN IIA—A Computer Code for Automated Common Cause Failure Analysis," TREE-1361 (May 1979).

7.5 ACTIVITY GENERATION, RELEASE AND TRANSPORT

Accident-Generated Activity Sources

The emphasis in the PRA analyses is on estimating the probabilities and frequency of occurrence of very rare events, such as a core meltdown followed by a breach of containment. In these circumstances a significant fraction of the volatile fission products could be released to the atmosphere, posing a hazard to the public. To analyze the complex mechanical, nuclear and chemical processes involved, a large effort, sponsored by NRC, EPRI and AIF, has been initiated to develop suitable computer codes and carry out their verification. The phenomena to be represented in the code equations are in part as follows:

1. Core Region: flow disturbances, fuel rod heatup and failure, coolant-clad chemical reactions, melting and slumping of fuel eutectics, molten pool formation and progression, radiative and convective heat transfer to other core components and coolant, debris bed cooling.
2. Vessel Head and Lower Plenum: heatup and failure of grid plates, melt component interactions, debris melting, heatup of vessel and its penetrations, vessel failure criteria.
3. Reactor Cavity and Concrete Basemat: concrete heatup and decomposition, chemical reaction with molten fuel and vessel, debris formation and sweepout, molten pool heat transfer, stratification and solidification.
4. Containment Building: intercompartment flow, heat transfer to structures, condensation and rainout of vapor-borne components, burning of combustible gases, steam explosions.

Several computer codes were developed to deal with the above phenomena in estimating the consequences of failure: MARCH, MELCOR, TRAC, CORCON and CONTAIN.[(1)–(5)]

Many of the safety codes discussed above are still undergoing development, with new features and capabilities added every year, so a detailed description of their internal organization is not warranted. Examples of codes developed to deal with the radionuclide buildup, release and transport phenomena are discussed below.

CORRAL Code

The CORRAL Code[6] was written at Battelle Columbus Laboratories to describe the transport of a nuclide within the containment building, after it has been released in an accident. It also deals with the phenomena occurring after the release of the radioisotope from the containment to the atmosphere. Code models were based on the results of containment systems experiments. The code was originally written to handle only light water reactor core failure, and the integral values for the release of fission products from molten fuel were based upon a number of very small experiments. The formulation of the code was later generalized to permit its use not only in the analysis of water reactor accidents, but also for liquid metal cooled breeders and gas-cooled reactors.

The capabilities of the code include the ability to take into account the natural transport of fission products and their deposition, or the prevention of their transport by the engineered safety features of the containment. Even though the source of the radioisotope release is assumed to be a core meltdown, less severe accidental releases can be treated as well. The model built into the code allows up to 15 individual compartments in the containment building through which the transport of radionuclides takes place. Up to 20 sequential events can be entered to describe the accident sequence. Fission products are handled in eight groups, and any of four release mechanisms can be considered.

The input requirements of the code include a description of the containment system, engineered safeguards systems, timing of events in the release sequence, thermodynamic conditions in the containment as a function of time, and the fission product release fractions. From these data the code computes the time rate of change in properties and the extent of fission product release. The set of differential equations describing fission product concentrations and accumulation in each compartment is solved incrementally, to obtain time-dependent cumulative fractional releases of activity from containment in each of the eight fission product groups.

The code is available for the CDC-7600 and CYBER73 machines. Running time is strongly dependent on the number of compartments specified for the interior of the containment building. For two compartment containment and 80 computational cycles the running time is 70 to 120 seconds on the CYBER73 machine. The same problem runs 1.2 seconds on the CDC-7600. A four-compartment problem with 100 computational cycles runs for 600 seconds on the CYBER73, and over 40 seconds on the CDC-7600. The code requires $137{,}000_8$ words of central processor memory and $52{,}000_8$ words of extended core memory.

Irradiation-Generated Activity Sources

ORIGEN Code of Oak Ridge

The ORIGEN Code[7] was written by M. J. Bell of Oak Ridge National Laboratory in 1973. A second version of the code was developed by A. G. Croff[8],[9] of Oak Ridge National Laboratory, and released in 1980. It is a versatile point depletion and decay computer code widely used in the simulation of nuclear fuel cycles and in calculations of the nuclear composition of irradiated fuels. The code had a worldwide distribution in its first version. In 1979 D. E. Bennett of Sandia National Laboratory released a version of ORIGEN with a much improved input and output capability.[10] In that same year a Japanese version of the code appeared.[11] An adaptation of the code for fusion work was carried out by J. A. Blink of Lawrence Livermore National Laboratory[12] and appeared in 1982. The latter version can accept more extensive activation cross sections, effective in the 1 to 20 MeV energy range.

The present capabilities of the code include the ability to calculate detailed isotopic compositions for a wide range of nuclear reactor concepts and fuel cycles. Nuclear accretion, activation and decay resulting from neutron irradiation are computed by solving an inhomogeneous system of coupled first-order differential equations describing allowable transitions. To make the use of the code easier in fuel reprocessing applications, the code contains provisions for entry of separation factors in the various reprocessing streams, which can be specified by element or element group. Chemical ingestion toxicity of the materials treated by the code can also be output.

The code is associated with extensive decay, cross section and photon libraries[13] which contain one-group, spectrum averaged neutron cross sections for 230 isotopes of actinides, fission products and structural materials developed from ENDF/B-IV and ENDF/B-V master libraries. The decay data contain half-lives, branching ratios, and recoverable energy per decay for 427 nuclides based on the ORNL Nuclear Structure Data File, and for 575 short-lived nuclides, based on ENDF/B-IV. The photon library is arranged in a 12-group structure for nonactinide isotopes, and an

18-group structure for actinides. It contains data for gamma rays, X rays and bremsstrahlung photons. Spectrum averaging of the multigroup cross sections and their collapse to one-group cross sections is done for each of the major reactor types for which appropriate material models were prepared. At present, ready-made models for PWR, BWR, LMFBR and CANDU reactor fuel cycles are available for input to the ORIGEN code.[(14)–(16)]

ORIGEN is available in several versions operating on different machines: CDC-6600, CDC-7600 and Cray-1. The source listing of the code and the corresponding report can be obtained from the Radiation Shielding Information Center at Oak Ridge, TN.

DKR Code of University of Wisconsin

The DKR Code[(17)] for radioactivity calculation was developed by Tak Yun Sung and William F. Vogelsang at the University of Wisconsin. The code is specifically intended for predicting activation in fusion devices. It computes the magnitude of alpha, beta and gamma ray sources in point, slab, cylinder, sphere and torus geometries. The following quantities are computed:

- Radioactivity
- Biological hazard potential
- Afterheat due to beta and gamma rays
- Decay gamma ray source

An ANISN module for transport theory calculation is part of the code and performs the prediction of distribution of photons from decay gamma ray sources. A DOSE auxiliary program is used to compute spatially dependent dose rates. A cross-section library is integral to the code and supplies data for the construction of linear decay chains which are then solved by a recursion coefficient method. Data for 29 transmutation chains are included, and are formulated in 46 neutron energy groups.

NAC Code of NASA

The NAC Code[(18)] was developed by Suzanne T. Weinstein of the NASA Lewis Center for the prediction of gamma ray activity induced by neutron capture. The code calculates activation induced either by steady state irradiation carried out for a fixed period and followed by a fixed period of decay, or through intermittent irradiation, where a specified number of irradiation and decay cycles are followed by a specified decay period.

The nuclear constants library built into the code contains four group neutron activation cross sections accounting for the gamma ray, proton, alpha and (n,2n) reactions and radioactive decay constants for decays into daughter and granddaughter products.

The input requires specification of the neutron group fluxes and weight fractions of the materials to be irradiated. The output gives a detailed breakdown of activation induced by gamma ray energy, and gamma ray dose rates due to it.

Airborne Radiation Hazards

Prediction of the magnitude of radiative hazards associated with inadvertent releases of fission products into the atmosphere can be obtained with computer codes such as EGAD[(19)] and EMERALD.[(20)] They compute the external gamma ray dose from airborne fission products. Radiation doses incurred by reactor personnel and the public at large, due to external exposure or ingested radioactive material, have been subject to extensive study and regulation.[(21)–(24)]

REFERENCES

1. R. O. Wooton and H. I. Avci, "MARCH Code Description and User's Manual," NUREG/CR-1711, BMI-2017 (Feb. 1970).
2. D. C. Aldrich and R. S. Denning, "MELCOR Program," LWR Safety Information Meeting, Gaithersburg, MD (Oct. 1982).
3. "TRAC-PD2—An Advanced Best-Estimate Computer Program for Pressurized Water Reactor Loss-of-Coolant Accident Analysis," NUREG/CR-2054, Los Alamos National Lab. (1982).
4. J. Muir et al., "CORCON-MOD1: An Improved Model for Molten Core/Concrete Interactions," NUREG/CR-2142, Nuclear Regulatory Commission, Washington, DC (1982).
5. M. E. Senglaub et al., "CONTAIN, A Computer Code for the Analysis of Containment Response to Reactor Accidents—Version 1," Draft NUREG/CR-2224, Sandia National Lab., Albuquerque, NM (1981).
6. R. J. Burian and P. Cybulskis, "CORRAL2 User's Manual," Battelle Columbus Lab. (Jan. 1977).
7. M. J. Bell, "ORIGEN: the ORNL Isotope Generation and Depletion Code," ORNL-4628, Oak Ridge National Lab. (May 1973).
8. A. G. Croff, "ORIGEN2—A Revised and Updated Version of the OAK Ridge Isotope Generation and Depletion Code," ORNL-5621, Oak Ridge National Lab. (1980).
9. A. G. Croff, "User's Manual for the ORIGEN2 Computer Code," ORNL/TM-7175, Oak Ridge National Lab. (1980).
10. David E. Bennett, "SANDIA-ORIGEN User's Manual," NUREG/CR-0987, Sandia National Lab., Albuquerque, NM (Dec. 1979).
11. K. Koyama, S. Miyasaka, and N. Yamano, "ORIGEN-JR: A Computer Code for Calculating Radiation Sources and Analyzing Nuclide Transmutations," JAERI M-8229, Japan Atomic Energy Research Inst., Tokyo, (May 1979).

12. J. A. Blink, "FORIG: A Modification of the ORIGEN2 Isotope Generation and Depletion Code for Fusion Problems," UCRL-53263, Lawrence Livermore National Lab. (Mar. 1982).
13. A. G. Croff, R. L. Haese, and N. B. Grove, "Updated Decay and Photon Libraries for the ORIGEN Code," ORNL/TM-6055, Oak Ridge National Lab. (Feb. 1979).
14. A. G. Croff, M. A. Bjerke, G. W. Morrison, and L. M. Petrie, "Revised Uranium–Plutonium Cycle PWR and BWR Models for the ORIGEN Computer Code," ORNL/TM-6051, Oak Ridge National Lab. (Sept. 1978).
15. A. G. Croff and J. W. McAdoo, "LMFBR Models for the ORIGEN Computer Code," ORNL/TM-7176, Oak Ridge National Lab. (Oct. 1981).
16. A. G. Croff and M. A. Bjerke, "CANDU Models for the ORIGEN Computer Code," ORNL/TM-7177, Oak Ridge National Lab. (Nov. 1980).
17. T. Y. Sung and William F. Vogelsang, "DKR: A Radioactivity Calculation Code for Fusion Reactors," VWFDM-170, Univ. Of Wisconsin, (Sept. 1976).
18. Suzanne T. Weinstein, "NAC: Neutron Activation Code," NASA-TM-X-52460, Lewis Research Center, Cleveland, OH (1968).
19. R. E. Cooper, "EGAD: A Computer Program to Compute Dose Integrals from External Gamma Emitters," DP-1304, Savannah River Lab. (Sept. 1972).
20. W. K. Brunot, R. R. Fray, and S. G. Gilespie "EMERALD (Revision 1) A Program for the Calculation of Activity Releases and Potential Doses from a Pressurized Water Reactor Plant," Pacific Gas and Electric Co. (Mar. 1974).
21. T. D. Luckey, *Hormesis with Ionizing Radiation,* CRC Press, Boca Raton, FL (1980).
22. UNSCEAR, "Sources and Effects of Ionizing Radiation," United Nations, New York (1977).
23. International Committee on Radiation Protection, *Permissible Dose for Internal Radiation,* ICRP Publication No. 2, Pergamon Press, New York (1959).
24. "Limits for Intakes of Radionuclides by Workers," International Committee on Radiation Protection, Publication No. 30, *Annals of ICRP,* 3, 1–4 (1980).

Nuclear Structure and Decay Schemes

The sources of man-made radiation met in practice consist of operating reactors or particle accelerators, products of neutron activation or fission products. The particulate and electromagnetic radiation produced has been the subject of many studies aimed at systematization and cataloging of radiation spectra, prediction of activity produced on irradiation and analysis of mechanisms of radiative source transport and deposition, as well as dosimetry. In these and similar studies, identification of the isotopes present requires knowledge of nuclear level schemes and the application of multichannel spectral analysis of the radioactive samples, as well as the ability to unfold from the radiation spectra the original spectrum responsible for the production of the radioactivants.

The evaluation and systematization of nuclear level schemes are an important preliminary step in the application of gamma ray spectral analysis. With the advent of reliable and reasonably priced multichannel analyzers, all endeavors relating to activation analysis have come into their own. Protection against ingestion of radioactivity borne by air, water and soil, plus an identification of the fission-product composition of spent fuels, both rely to some extent on the methodology of multichannel gamma spectrum analysis. The application of activation analysis requires, in turn, easy availability of information on nuclear decay schemes, radiation intensities, energies, branching ratios and decay half-lives. Compilations containing this type of data, in either tabular or graph form, appeared early in the history of the nuclear industry, and have been reissued many times.

One of the most useful reference tabulations of isotope data of this kind was originally published in 1940, in *Reviews of Modern Physics,* by J. J. Livingood and G. T. Seaborg under the title "Table of Isotopes."[(1)] Republished by John Wiley & Sons in a much expanded book format,[(2)] with Isadore Perlman as its first editor, it is now available in its seventh edition, and has four co-editors. In its latest form, the first part of the compilation contains isotope identification, decay half-life, type of decay and relative intensity, identification of generic relationships, energies (in MeV) of major radiations and the principal means of production. The second part of the publication contains graphs of decay schemes. The appendixes contain nuclear spectrographic standards, atomic energy levels with particular reference to K shell X-ray data, nuclear decay rates and theoretical nuclear level diagrams.

The earlier editions of the above tables selectively concentrated on the prominent gamma rays associated with a substance, and ignored the bulk of emissions of very low intensity as being of little interest to the practicing nuclear engineer, chemist or activation scientist. They took cognizance of the fact that such readers are usually looking for easily identifiable radiation. It must be observed, however, that the latest (seventh) edition of *Table of Isotopes* followed the direction of the primary reference in the field, *Nuclear Data Sheets*. In fact, it uses the same format in its nuclear decay schemes as *Nuclear Data Sheets,* thereby adding much detail, but losing in the process part of its earlier appeal of simplicity and ease of interpretation.

As was stated above, the primary reference data collection for nuclear decay schemes is contained in *Nuclear Data Sheets*. Initially published in a loose-leaf form, and presenting decay schemes plotted by hand, the sheets are now generated, revised and plotted by machine, as part of the Nuclear Data Project at ORNL. For every known nucleus, the Evaluated Nuclear Structure Data File (ENSDF) of ORNL contains information used to generate *Nuclear Data Sheets*.[(3)] The same data file can be used to extract specific nuclear properties, or particular types of experimental information.

Input data for the file are either abstracted from papers published by experimenters in successive issues of *Nuclear Physics,* or entered into the file directly from information supplied by the experimenters. To facilitate this process, the Nuclear Data Project publishes a manual[(4)] for the preparation of data sets.

Associated with ENSDF is a set of servicing computer programs containing the following components:

- NDSLIST: makes up tables for publication in *Nuclear Data Sheets.*
- PLOT: produces drawings for *Nuclear Data Sheets.*
- LOGFT: calculates logarithms of ft's and shell conversion energy.
- ALPHAHF: calculates alpha-decay hindrance factors.
- GTOL: calculates best level energies.
- HSICC: calculates total conversion coefficients.

The data sheets are no longer published in loose-leaf form. Instead, the task of publishing them was taken over in 1973 by the Academic Press of New York, and they are issued in soft-cover books in 8 1/2″ by 11″ format for easier distribution. This has now become a multivolume set of imposing dimensions.

REFERENCES

1. J. J. Livingood and G. S. Seaborg, "Table of Isotopes," *Reviews of Modern Physics,* 12, 30 (1940).
2. C. Michael Lederer, Jack M. Hollander, and Isadore Perlman, *Table of Isotopes,* 6th edit., John Wiley & Sons, New York (1967).
3. *Nuclear Data Sheets,* Academic Press, New York (published from 1973 onward).
4. W. Bruce Ewbank and Marcel R. Schmorak, "Evaluated Nuclear Structure Data File, A Manual for Preparation of Data Sets," ORNL-5054/R1, Oak Ridge National Lab. (Feb. 1978).

Spectral Unfolding Codes and Standards

Neutron and gamma ray spectra generated by reactors in activated materials and fission products are available in several tabulations.[(1)–(4)] The components contributing to their structure can be obtained by spectral unfolding from the response of dynamic or activation detectors. The spectral unfolding is now mostly handled by either of two numerical techniques. The matrix inversion method obtains the unfolded spectrum from the inverted product of the experimentally measured or calculated response matrices and the detector readings. The derivative method uses an idealized approximation of the response functions resembling actual distributions. These methods are applicable in the unfolding of gamma ray spectra, as well as the spectra obtained from proton recoil spectrometers, and from activation or threshold detectors. A Monte Carlo approach to spectral unfolding has been also attempted.[(5)]

A number of spectrum unfolding codes have been written and are now in use.[(6),(7)] Among the matrix inversion group the FERDOR and MINREP codes[(8),(9)] are representative of those most often used, while FLYSPEC and MATXUF[(10),(11)] are the leading contenders in the derivative group.

Differences in the mathematical approach and in the response of the detectors used led to difficulty in the intercomparison of unfolded results. To bring some coherence to this field, an effort was initiated by Walter Meyer under the auspices of the ANS 6.2.2 Spectra Unfolding Working Group to prepare standards for the performance of the spectrum unfolding codes. A similar effort was begun by the ASTM E10.05 Subcommittee on Dosimetry.

The ANS standardization effort is seen as basically a three-step process whereby, first, the unfolding codes are tested using idealized, calculated response functions on synthesized spectra; second, the codes are tested using measured, smoothed response functions on synthesized spectra; and, finally, both of the preceding response functions are used on experimentally measured spectra.

The contribution of the development of experimental neutron activation analysis and gamma-ray spectrometry to advancing the state of art in the area of computerized spectral unfolding is obvious. Progress in this area was highlighted by a topical conference on the use of computers in activation analysis and gamma-ray spectroscopy, held in May of 1978 in Mayaguez, Puerto Rico, under the auspices of ANS.

REFERENCES

1. R. E. Maerker and F. J. Muckenthaler, "Gamma-Ray Spectra Arising from Fast-Neutron Interactions in Elements Found in Soils, Concretes and Structural Materials," ORNL-4475, Oak Ridge National Lab. (1970).
2. R. E. Maerker and F. J. Muckenthaler, "Gamma-Ray spectra Arising from Thermal-Neutron Capture in Elements Found in Soils, Concretes, and Structural Materials," ORNL-4382, Oak Ridge National Lab. (1970).
3. L. Celnik and D. Spielberg, "Gamma Spectral Data for Shielding and Heating Calculations," NASA CR-54794, UNC-5140, United Nuclear Co., Elmsford, NY (1965).
4. C. Z. Serpan and B. H. Menke, "Nuclear Reactor Neutron Energy Spectra," DS52, American Society for Testing Materials (1974).
5. Robert S. Sanna, "A User's Guide to SWIFT, a Monte Carlo Technique for Unfolding Neutron Spectra," HASL-224, U.S. Atomic Energy Commission (July 1971).

6. P. J. M. Korthoven, "RESOLF—A Computer Program for the Analysis of Gamma-Ray Spectra," IS-1811, Iowa State University, Ames Lab. (1968).
7. F. B. K. Kam and F. W. Stallman, "CRYSTAL BALL: A Computer Program for Determining Neutron Spectra from Activation Measurements," ORNL-TM-4601 (June 1974).
8. Walter R. Burrus, "FERD and FERDOR Type Unfolding Codes," in: "A Review of Radiation Energy Spectra Unfolding," compiled by D. K. Trubey, ORNL/RSIC-40, pp. 1–22 (Oct. 1976).
9. P. G. Kirmser, K. K. Hu, and W. Meyer, "A New Method for Solving the Unfolding Problem for Bounded, Inaccurate Radiation Data," in "A Review of Radiation Energy Spectra Unfolding," compiled by D. K. Trubey, ORNL/RSIC-40, pp. 107–116 (Oct. 1976) (MINREP code).
10. D. R. Slaughter, "A Simple Method to Unfold NE-213 and Stilbene Neutron Spectral Data," in: "A Review of Radiation Energy Spectra Unfolding," compiled by D. K. Trubey, ORNL/RSIC-40, pp. 65–77 (Oct. 1976) (FLYSPEC code).
11. W. H. Miller, "MATXUF: On-line, Real Time Derivative Method Neutron Spectrum Unfolding," *Nucl. Instr. & Methods*, 153, 535–541 (July 1978).

8. ENGINEERING SOCIETY SERIALS AND STANDARDS

8.1 JOURNALS AND STANDARDS OF ENGINEERING SOCIETIES

As the discipline of nuclear engineering was maturing, several engineering societies active in this field started publication of specialist journals and set up committees charged with the establishment of performance and quality standards for the components, operating procedures and computer codes used for nuclear reactors.

The engineering standard is a prescription for a product, process or procedure for testing or manufacturing, about which a consensus has been reached among knowledgeable practitioners in the field that it represents an acceptable practice, quality or performance. Once its text has been formulated and approved by generally accepted procedures, it is published for use in industrial practice. The published document can vary in length from a few pages to several thousand, but regardless of its length each of these documents contains a description of the situation creating the need for the standard, a detailed prescription for the process or product that if followed will produce the required result, and, finally, provisions for its periodic review.

Standardization Activities of ANSI

The society standards committees usually operate within the framework of the American National Standards Institute (1430 Broadway, New York, NY 10018), and the standards developed often bear the double designation of ANSI and the issuing society. Once created, the standards are used by reactor designers and manufacturers, suppliers of nuclear components, architect-engineer firms, construction companies, utilities operating nuclear plants, and the personnel of regulatory bodies. In recent times the Nuclear Regulatory Commission has quoted in its Regulatory Guides several of the existing standards, thus setting the level of quality and performance to be aimed at in applications for plant licenses. This points to the growing importance of standards in the licensing area.

The American National Standards Institute acquired its present name in 1969, as the most recent step in its growth, which started in 1918 with the formation of the American Engineering Standards Corporation. Transformed into the American Standards Association in 1928, and USA Standards Institute in 1966, it is the body responsible for overseeing the process of establishing standards by consensus of all parties concerned. ANSI does not write standards directly but concentrates on securing the evidence of consensus for each one. Consensus can be arrived at in one of three acceptable methods involving accredited bodies: (1) organization method, (2) standards committee method, (3) canvass method.

Support for the development of nuclear standards constitutes just one aspect of ANSI's activities, and its interactions with the various engineering societies active in the nuclear field have been consolidated under the auspices of the Nuclear Standards Management Board (NSMB), which meets several times a year. Its functions include identification and definition of the need for a standard, assignment of responsibility for its development to one of the accredited organizations, monitoring the performance of the developing organization, coordination of development activities, and assurance that one of the three methods of arriving at a consensus for a standard has been in fact achieved.

The ANSI procedure for the development of nuclear standards involves the following steps:

- A scope for the standard is developed.
- A project charter is prepared.
- Drafts are developed by working groups.
- Drafts are reviewed by the NSMB Planning Committee.
- The standard is approved by NSMB.

Drafts must be approved in turn by the working groups, subcommittees and finally the main committee. Should a consensus be lacking in the main committee, the project is dropped. While the standard is in the proposal stage, public review and comment are solicited.

Granting approval of a standard by ANSI is based on the ability of the organization and the committee submitting the standard to demonstrate that all substantially affected interests have been given an opportunity to participate in its formulation, and to comment on its final form. Formal submission of evidence of consensus to the ANSI Board of Standards Reviews is required before final approval is obtained. Publication follows approval.

The above procedure may be reversed in case of obsolete standards, and standards may be withdrawn, or replaced by an updated version. A conscious effort is made to review and, if necessary, revise the existing standards on a five year schedule to assure that the standards reflect the current state of the art.

A complete status report[1] of nuclear standards already approved, under revision or proposed is published quarterly in February, May, August and November of each year, and may be obtained by subscription from NSMS for an annual fee of $100. Proposed standards are advertised in the biweekly *ANSI Reporter,* and public response is invited from interested persons during the 60-day comment period.

The standards covered here are those that bear directly on the field of nuclear engineering; they form a subset of a much broader category of national standards. Included in the discussion will be the nuclear standards developed by the American Nuclear Society, American Association of Mechanical Engineers, Institute of Electrical and Electronics Engineers, American Society for Testing and Materials, National Bureau of Standards, Health Physics Society, Institute of Nuclear Materials Management, Instrument Society of America, American Institute of Chemical Engineers, and National Fire Protection Association.

The standards in this group relate to the manufacture, construction and erection of nuclear power stations, the performance of items of equipment incorporated in these stations, the instrumentation used in operating the plants and research reactors, and the protection of personnel and the general public from radiation exposure. Also covered by these standards are the processes involved in the mining and purification of actinide fuels, the manufacture and transport of new fuel elements, and the storage and reprocessing of spent fuel, as well as the question of safe storage of spent fuel and radioactive wastes.

A separate area of standards, very important to the field of nuclear numerical analysis, deals with the quality, format and performance of computing equipment, with the magnetic recording media as well as with the transmission of numerical and textual information over the public communication networks. Only a small part of standards in this area are included in the activities of the Nuclear Standards Board, namely the standards for writing nuclear computer codes; others fall under the category of information transfer. The relevant standards in the information area are listed in the following section.

REFERENCES

1. "ANSI-NSB Status Report of Nuclear Standards," American National Standards Institute, New York. Issued quarterly.

Listing of Standards for Information Recording and Transfer

	COMPUTERS
ANSI/ASTM E731-80	Guidelines for the Procurement of Commercially Available Computerized Systems.
ANSI/ASTM E730-80	Guidelines for Developing Functional Designs for Computerized Systems.
ANSI/NFPAM75-1981	Protection of Electronic Computer Data Processing Equipment.
	COMPUTER PROGRAMMING
ANSI/ANS 10.2-1982	Recommended Programming Practices to Facilitate Portability of Scientific Computer Programs.
ANSI/ANS 10.5-1979	Guidelines for Considering User Needs in Computer Program Development.
ANSI/IEEE 730-1981	Software Quality Assurance Plans.
ANSI X3.37-1980	Programming Language APT.
ANSI X3.23-1974	Programming Language COBOL.
ANSI X3.9-1978	Programming Language FORTRAN.
ANSI X3.53-1976	Programming Language PL/I.
ANSI X3.60-1978	Minimal BASIC.
ANSI/ ISA S61.1-1976	Industrial Computer System FORTRAN Procedures for Executive Functions, Process Input/Output, and Bit Manipulation.
ANSI/ ISA S61.2-1978	Industrial Computer System FORTRAN Procedures for File Access and Control of File Content.
	MAGNETIC RECORDING MEDIUM
ANSI X3.40-1976	Unrecorded Magnetic Tape for Information Interchange (9-Track 200 and 800 CPI, NRZI, and 1600 CPI, PE).
ANSI X3.76-1981	General, Physical and Magnetic Requirements, Unformatted Single Disk Cartridge (Top Loading, 200 TPI, 4400 BPI).
ANSI X3.84-1981	General, Physical and Magnetic Requirements, Unformatted Twelve-Disk Pack (200 Megabytes).
ANSI X3.82-1980	One-Sided, Single-Density Unformatted 5.25 Inch Flexible Disk Cartridge.
ANSI X3.27-1978	Magnetic Tape Labels and File Structure for Information Interchange.
	DATA TRANSFER BY TELECOMMUNICATION NETWORKS
ANSI X3.41-1974	Code Extension Techniques for Use with the 7-Bit Coded Character Set of American National Standard Code for Information Interchange (ASCII).
ANSI X3.28-1976	Procedures for the Use of Communications Control Characters of American National Standard Code for Information Interchange in Specified Data Communication Links.

Listing of Standards for Information Recording and Transfer (Continued)

ANSI X3.4-1977	Code for Information Interchange.
ANSI X3.42-1975	Representation of Numeric Values in Character Strings for Information Interchange.
ANSI X3.36-1975	Synchronous High-Speed Data Signalling Rates Between Data Terminal Equipment and Data Communication Equipment.
ANSI/IIA RS-404-1978	Start–Stop Signal Quality Between Data Terminal Equipment and Non-Synchronous Data Communication Equipment.

Serials Published by ANS

Several engineering societies are active in publishing periodical literature in the field of nuclear engineering. Foremost among them is the American Nuclear Society, with headquarters at 555 North Kensington Ave., La Grange Park, IL 60525.

Since 1956 the American Nuclear Society has published a monthly journal, *Nuclear Science and Engineering,* to present fundamental research in the area of nuclear sciences and engineering. *The Nuclear News* also appears on a monthly basis. The activities of members of ANS in the area of technology were initially covered by the monthly journal *Nuclear Applications,* later renamed *Nuclear Technology* and published jointly with the European Nuclear Society. Since 1981 *Nuclear Technology* has also appeared in the *Fusion* version. In July 1984 *Nuclear Technology/Fusion* was renamed *Fusion Technology* and appears as a separate technical journal. *Transactions of the American Nuclear Society* is published biannually, to correspond with the June and November meetings of the society, and is the main vehicle for bringing to the attention of ANS members the latest developments in the field of nuclear engineering. In addition, in the February and August issues of *Nuclear News,* ANS publishes a list of nuclear plants, with power rating of 30 MWe or over, operating, under construction, or on order as of June 30 and December 31 of each year. A monograph series dealing with nuclear subjects is also published.

The Society's periodicals are as follows:

Title	Volume & Period
• *Transactions*	1–, 1958–
• *Nuclear Applications*	1–6, 1965–1969
• *Nuclear Applications and Technology* (formerly: *Nuclear Applications*)	7–9, 1969–1970

Title	Volume & Period
• *Nuclear News*	1–, 1959–
• *Nuclear Report*	1–, 1978–
• *Nuclear Science and Engineering*	1–, 1956–
• *Nuclear Standards News*	1–, 1971–
• *Nuclear Technology,* (formerly: *Nuclear Applications and Technology*)	10–, 1971–
• *Nuclear Technology/Fusion*	1–, 1981–
• *Fusion Technology* (formerly: *Nuclear Technology/Fusion*)	5–, 1984–
• *Remote Systems Technology Proceedings*	annual

Development of Standards by ANS

As might be expected, the membership of the American Nuclear Society is extensively involved in the development of standards for the nuclear industry. ANS standards activities began in 1957, and its standards committee now numbers some 1320 members. Table 1 summarizes the growth of membership of the committee and the number of standards produced each year.[1]

The actual writing of proposed standards is done by the working groups, of which there are now some 150 within ANS. Their membership is voluntary and is recruited from professionals with interest in the particular subject and technical expertise in the field. The working groups operate within 20 subcommittees responsible for conducting peer reviews of proposed standards to obtain technical approval. Once this approval is obtained, the proposed standard is placed in the hands of one of the four consensus committees for additional peer review and formal balloting. The consensus committees are responsible for: (1) nuclear power plant standards; (2) research reactors, reactor physics, and radiation shielding; (3) nuclear criticality safety; (4) radioactive waste management. Overall

Table 1. Membership and Production of ANS Standards Committee

Year	Membership No.	Standards	Year	Membership No.	Standards
1972	250	10	1978	1079	15
1973	600	2	1979	1133	21
1974	700	6	1980	1180	6
1975	800	10	1981	1235	13
1976	900	15	1982	1260	13
1977	1042	9	1983	1320	10

direction of these four groups is provided by the ANS Standards Steering Committee.

In the preparation of standards, the standards committee of ANS decided in 1976 to adopt the terminology of nuclear science outlined in the glossary[2] published by ANS and approved as National Standard N1.1-1967 by the American National Standards Institute. It was hoped that this action would promote the use, by the members of the nuclear industry at large, of the same common language of technical communication.

The ANS standards are categorized under the following headings: nuclear terminology and units, radioactive materials handling, nuclear criticality safety, reactor physics and shielding, nuclear power plant standards, nuclear design criteria, radiological environmental effects, and, radioactive waste management. The largest of these categories is the one for power plant standards.

The full text of standards issued by the ANS up to 1977 was published under one cover in that same year,[3] together with a review of ANS standards activities. ANS standards approved by the American National Standards Institute are identified in this publication as American National Standards.

The titles of ANS standards are listed below. Copies of the full text of individual standards may be purchased directly from ANS (555 North Kensington Ave., La Grange Park, IL 60525) at prices varying between $5.50 and $55.00, depending on the length of the document. In addition, in 1971 ANS created the Information Center on Nuclear Standards (ICONS) to serve as a clearinghouse for up-to-date information on all nuclear standards, including those developed by other engineering societies. *Nuclear Standards News* serves as the information medium for ICONS.

REFERENCES

1. Marilyn D. Weber, "Nuclear Standards Development in the U.S.—How ANS Fits In," ANS News, 26, 5–6 (Nov. 1983).
2. "Glossary of Terms in Nuclear Science and Technology," American Nuclear Society, La Grange Park, IL (1976).
3. "American Nuclear Society Standards, 1977 Edition," American Nuclear Society, La Grange Park, IL (1977).

Listing of Nuclear Standards Issued by ANS

	CRITICAL EXPERIMENTS
ANS-1-1975(81)	Safety Guide for the Performance of Critical Experiments.
	NUCLEAR POWER PLANT SAFETY
ANS-2.2-1978	Earthquake Instrumentation Criteria for Nuclear Power Plants.
ANS-2.3-1979	Immediate Evacuation Signal for Use in Industrial Installations.
ANS-2.7-1982	Guidelines for Assessing Capability for Surface Faulting at Nuclear Power Reactor Sites.
ANS-2.8-1981	Determining Design Basis Flooding at Power Reactor Sites.
ANS-2.9-1980	Evaluation of Ground Water Supply for Nuclear Power Sites.
ANS-2.10-1979	Guidelines for Retrieval, Review, Processing and Evaluation of Records Obtained from Seismic Instrumentation.
ANS-2.11-1978	Guidelines for Evaluating Site-Related Geotechnical Parameters at Nuclear Power Sites.
ANS-2.12-1978	Guidelines for Combining Natural and External Man-Made Hazards at Power Reactor Sites.
ANS-2.13-1979	Evaluation of Surface-Water Supplies for Nuclear Power Sites.
ANS-2.17-1980	Evaluation of Radionuclide Transport in Ground Water for Nuclear Power Sites.
ANS-2.19-1981	Guidelines for Establishing Site-related Parameters for Site Selection and Design of an Independent Spent Fuel Storage Installation (Water Pool Type).
ANS-3.1-1981	Selection, Qualification and Training of Personnel for Nuclear Power Plants.
ANS-3.2-1982	Administrative Controls and Quality Assurance for the Operational Phase of Nuclear Power Plants.
ANS-3.3-1982	Security for Nuclear Power Plants.
ANS-3.5-1981	Nuclear Power Plant Simulators for Use in Operator Training.
ANS-3.7.1-1979	Facilities and Medical Care for On-Site Nuclear Power Plant Radiological Emergencies.
ANS-3.7.2-1979	Emergency Control Centers for Nuclear Power Plants.
ANS-3.7.3-1979	Radiological Emergency Preparedness for Nuclear Power Plants.
ANS-4.5-1980	Criteria for Accident Monitoring Functions in Light-Water-Cooled Reactors.
ANS-5.1-1979	Decay Heat Power in Light Water Reactors.
	RADIATION SHIELDING
ANS-6.1.1-1977	Neutron and Gamma-Ray Flux-to-Dose-Rate Factors.
ANS-6.3.1-1980	Program for Testing Radiation Shields in LWR's.
ANS-6.4-1977	Guidelines on the Nuclear Analysis and Design of Concrete Radiation Shielding for Nuclear Power Plants.
ANS-6.6.1-1979	Calculation and Measurement of Direct and Scattered Gamma Radiation from LWR Nuclear Power Plants.
ANS/HPSSC-6.8.1-1981	Location and Design Criteria for Area Radiation Monitoring Systems for Light Water Nuclear Reactors.

Listing of Nuclear Standards Issued ANS (Continued)

Standard	Title
IEEE/ ANS-7- 4.3.2-1981	Application Criteria for Programmable Digital Computer Systems in Safety Systems of Nuclear Power Generating Stations.
	NUCLEAR CRITICALITY SAFETY
ANS-8.3-1979	Criticality Accident Alarm System.
ANS-8.5-1979	Use of Borosilicate-Glass Raschig Rings as a Neutron Absorber in Solutions of Fissile Materials.
ANS-8.7-1975(82)	Guide for Nuclear Criticality Safety in the Storage of Fissile Materials.
ANS-8.9-1978	Nuclear Criticality Safety Guide for Pipe Intersections Containing Aqueous Solutions of Enriched Uranyl Nitrate.
ANS-8.12-1979	Nuclear Criticality Control and Safety of Homogeneous Plutonium–Uranium Fuel Mixtures Outside Reactors.
ANS-8.15-1981	Nuclear Criticality Control of Special Actinide Elements.
	APPLICATION OF COMPUTERS
ANS-10.2-1982	Recommended Programming Practices to Facilitate the Portability of Scientific Computer Programs.
ANS-10.5-1979	Guidelines for Considering User Needs in Computer Program Development.
ANS-14.1-1982	Operation of Fast Pulse Reactors.
	RESEARCH REACTORS
ANS-15.1-1982	Development of Technical Specifications for Research Reactors.
ANS-15.4-1977	Selection and Training of Personnel for Research Reactors.
ANS-15.7-1977	Research Reactor Site Evaluation.
N402-1976 (ANS-15.8)	Quality Assurance Program Requirements for Research Reactors.
ANS-15.10-1981	Decommissioning of Research Reactors.
ANS-15.11-1977	Radiological Control at Research Reactor Facilities.
ANS-15.12-1977	Design Objectives for and Monitoring of Systems Controlling Research Reactor Effluents.
ANS-15.15-1978	Criteria for the Reactor Safety Systems of Research Reactors.
ANS-15.16-1982	Emergency Planning for Research Reactors.
ANS-15.17-1981	Fire Protection Program Criteria for Research Reactors.
	PLANT DESIGN
ANS-18.5-1982	Surveys of Terrestrial Ecology Needed to License Thermal Power Plants.
ANS-19.5-1978	Requirements for Reference Reactor Physics Measurements.
ANS-51.1-1975 (N18.2a-1975)	Nuclear Safety Criteria for the Design of Stationary PWR Plants.
ANS-51.10-1979	Auxiliary Feedwater System for PWR's.
ANS-55.1-1979	Solid Radioactive Waste Processing System for Light Water Cooled Reactor Plants.

Listing of Nuclear Standards Issued by ANS (Continued)

ANS-55.4-1979	Gaseous Radioactive Waste Processing Systems for Light Water Reactor Plants.
ANS-55.6-1979	Liquid Radioactive Waste Processing System for Light Water Reactor Plants.
ANS-56.5-1979	PWR and BWR Containment Spray System Design Criteria.
ANS-56.6-1978	PWR Containment Ventilation Systems.
ANS-56.7-1978	BWR Containment Ventilation Systems.
ANS-56.8-1981	Containment System Leakage Testing Requirements.
ANS-56.10-1982	Subcompartment Pressure and Temperature Transient Analysis in LWRs.
ANS-57.1-1980	Design Requirements for Light Water Reactor Fuel Handling Systems.
ANS-57.5-1981	Light Water Reactors Fuel Assembly Mechanical Design and Evaluation.
ANS-57.7-1981	Design Criteria for an Independent Spent Fuel Storage Installation (Water Pool Type).
ANS-57.8-1978	Fuel Assembly Identification.
ANS-58.2-1980	Design Basis for Protection of Light Water Nuclear Power Plants Against Effects of Postulated Pipe Rupture.
ANS-58.3-1977	Physical Protection for Systems and Components Important to Safety.
ANS-58.4-1979	Criteria for Technical Specifications for Nuclear Power Stations.
ANS-58.9-1981	Single Failure Criteria for LWR Safety-Related Fluid Systems.
ANS-59.1-1979	Safety Related Cooling Water Systems in Nuclear Power Plants.
ANS-59.4-1979	Generic Requirements for Light Water Nuclear Power Plant Fire Protection.

Serials Published by ASME

The American Society of Mechanical Engineers, with headquarters at 345 East 47th Street, New York, NY 10017, is involved in the promotion and development of knowledge in the field of mechanical devices, pressure vessels, energy conversion equipment, heat transfer, fluid flow, strength of materials, and energy economics. The journals published by the Society reflect its interests, and in several fields overlap areas of interest of nuclear engineering.

Nuclear-related ASME publications are:

Title	Volume & Period
• *Journal of Applied Mechanics*	81–, 1959–
• *Journal of Energy Resources*	101–, 1979–
• *Journal of Fluids Engineering*	95–, 1973–
• *Journal of Heat Transfer*	88–, 1959–
• *Journal of Pressure Vessel Technology*	96–, 1974–

Title	Volume & Period
• Boiler and Pressure Vessel Codes	1974, 1983
• Companies Holding Nuclear Certificates of Authorization	1975

Development of Standards by ASME

The second most active participant in the nuclear standards field is the American Society of Mechanical Engineers. The development of nuclear standards of ASME falls within the area of competence of its Board of Nuclear Codes and Standards. Standards activities are carried out by the following committees:

- Nuclear air and gas treatment equipment.
- Concrete reactor and containment vessels (jointly with ACI).
- Nuclear cranes.
- Nuclear quality assurance.
- Nuclear inservice inspection.
- Operation and maintenance.
- Qualification testing for functional adequacy of nuclear plant.
- Qualifications for nuclear inspection and professional engineers.
- Nuclear power.

The eight standards in the N45.2 series are currently endorsed as such in Nuclear Regulatory Guide 1.28. They have recently been combined by ASME into one document: NQA-1-1979, "Quality Assurance Program Requirements for Nuclear Power Plants" (see listing below). There is a proposed revision to Nuclear Regulatory Guide 1.28 that would take cognizance of the combined standard NQA-1-1979, in lieu of the N45.2 standard series currently endorsed.

Building on the experience accumulated during the past seven decades in the preparation of the Boiler and Pressure Vessel Code, ASME has produced additional standards for large nuclear components by extending the sections of that code dealing with unfired pressure vessels and piping. First formulated by the society in 1914, the Boiler and Pressure Vessel Code has undergone many revisions and extensions.(1) The code is kept up to date by a volunteer group of more than 960 engineers, organized into the Boiler and Pressure Vessel Committee, and into several subcommittees. In part, or in its entirety, the code is now adopted in the United States by 46 states, and in all the provinces of Canada. The code has two specifically nuclear sections: power plant components and containment vessels in Section III, and nuclear in-service inspection, in Section XI.

The importance of the pressure vessel committee is attested to by the growth in the size of the corresponding section of the Pressure Vessel and Boiler Code produced under its direction. Code Section III, dealing with the construction of nuclear power plant components, consists now of eight volumes. Originally issued on November 4, 1963, it is currently being revised and updated twice a year. Its subsection NB, dealing with Class 1 components, up to and including the winter of 1981 addenda, has been adopted as part of the Federal Code of Regulations, 10 CFR 50. Subsections NC and ND, dealing with Class 2 and 3 components, were proposed for inclusion in 10 CFR 50 by the Nuclear Regulatory Commission. Code Section XI, Rules for Inservice Inspection of Nuclear Power Plant Components, was first issued in 1970 and is now being revised annually.

The entire Boiler and Pressure Vessel Code consists of 11 volumes comprising some 8500 pages, and sells for $3,000 to nonmembers and $2,400 to ASME members. Starting in July 1983, the English unit version of the code was issued in loose-leaf form only, with the price including the update service. The metric version of the code in SI units, made available in October 1983, provides all material in metric units only, including all equations, charts, graphs and tables of material properties. In addition to the code proper, interpretations of the code and sample analyses for nuclear piping systems are also available to users.[(2),(3)]

Those companies that manufacture their nuclear equipment according to ASME nuclear quality standards and pass appropriate inspection become holders of a Nuclear Certificate of Authorization, permitting them to stamp their products with the Nuclear Quality Stamp. A bimonthly list of such companies is published by ASME.[(4),(5)]

REFERENCES

1. "History of the ASME Boiler Code," American Society of Mechanical Engineers, New York (1955).
2. "Interpretations of ASME Boiler and Pressure Vessel Code," American Society of Mechanical Engineers, New York (1983).
3. "Sample Analysis of A Piping System Class 1 Nuclear," American Society of Mechanical Engineers, New York (1972).
4. "Companies Holding Nuclear Certificates of Authorization," American Society of Mechanical Engineers, New York (1983).
5. "Companies Holding Boiler and Pressure Vessel Certificates of Authorization for Use of Code Symbol Stamps," American Society of Mechanical Engineers, New York (1983).

Listing of Nuclear Standards Issued by ASME

NUCLEAR QUALITY ASSURANCE	
N45.2/N46.2-1977 (see NQA-1-1979)	Quality Assurance Program Requirements for Nuclear Facilities.
N45.2.1-1980	Cleaning of Fluid Systems and Associated Components for Nuclear Power Plants.
N45.2.2-1978	Packaging, Shipping, Receiving, Storage and Handling of Items for Nuclear Power Plants.
N45.2.3-1973 N45.2.3a-1978	Housekeeping During the Construction Phase of Nuclear Power Plants.
N45.2.5.-1978	Supplementary Quality Assurance Requirements for Installation, Inspection, and Testing of Structural Concrete and Structural Steel, Soils and Foundations During the Construction Phase of Nuclear Power Plants.
N45.2.6-1978 (see NQA-1-1979)	Qualifications of Inspection, Examination, and Testing Personnel for the Construction Phase of Nuclear Power Plants.
N45.2.8-1975 (R1980) N45.2.8a-1981	Supplementary Quality Assurance Requirements for Installation, Inspection, and Testing of Mechanical Equipment and Systems for the Construction Phase of Nuclear Power Plants.
N45.2.9-1979 (see NQA-1-1979)	Requirements for Collection, Storage, and Maintenance of Quality Assurance Records for Nuclear Power Plants.
N45.2.10-1979 (see NQA-1-1979)	Quality Assurance Terms and Definitions.
N45.2.11-1979 (see NQA-1-1979)	Quality Assurance Requirements for the Design of Nuclear Power Plants.
N45.2.12-1977 (see NQA-1-1979)	Requirements for Auditing of Quality Assurance Programs for Nuclear Power Plants.
N45.2.13-1979 (see NQA-1-1979)	Quality Assurance Requirements for Control of Procurement of Items and Services for Nuclear Power Plants.
N45.2.15-1981	Hoisting, Rigging, and Transporting of Items for Nuclear Power Plants.
N45.2.20-1979	Supplementary Quality Assurance Requirements for Subsurface Investigation Prior to the Construction Phase of Nuclear Power Plants.
N45.2.23-1978 (see NQA-1-1979)	Qualification of Quality Assurance Program Audit Personnel for Nuclear Facilities.
NQA-1-1979	Quality Assurance Program Requirements for Nuclear Power Plants.
NQA-2-1983	Quality Assurance Requirements for Manufacturing and Construction for Nuclear Power Plants.

INSERVICE INSPECTION	
ANSI/ASME B & PVC-1980	Section XI, Rules for Inservice Inspection of Nuclear Power Plants Components (Division 1).

Listing of Nuclear Standards Issued by ASME (Continued)

	QUALIFICATIONS AND DUTIES
N626.0-1982	Qualifications and Duties for Authorized Inspection Agencies, Inspectors and Nuclear Inspector Supervisors.
N626.1-1982	Qualifications and Duties for Authorized Inspection Agencies, Nuclear Inservice Inspectors and Nuclear Inservice Inspector Supervisors.
N626.2-1982	Qualifications and Duties for Authorized Nuclear Inspection Agencies, Inspectors (Concrete) and Nuclear Inspector Supervisors (Concrete).
N626.3-1979	Qualifications and Duties of Personnel Engaged in ASME Boiler and Pressure Vessel Code, Section III, Division 1 and 2, Certifying Activities.
	OPERATION AND MAINTENANCE
OM-1-1981	Requirements for Inservice Performance Testing of Nuclear Power Plant Pressure Relief Devices.
OM-2-1982	Requirements for Performance Testing of Nuclear Power Plant Closed Water Systems.
OM-3-1982	Requirements for Preoperational and Initial Start-up Vibration Testing of Nuclear Power Plant Piping Systems.
OM-4-1982	Examination and Performance Testing of Nuclear Power Plant Dynamic Restraints (Snubbers).
OM-5-1981	Inservice Monitoring of Core Support Barrel Axial Preload in Pressurized Water Reactors.
	PERFORMANCE TEST CODES
ANSI/ASME PTC 32.1-1974	Nuclear Steam Supply Systems.
ANSI/ASME PTC 32.2-1978	Methods of Measuring the Performance of Nuclear Reactor Fuel in Light Water Reactors.
	NUCLEAR AIR AND GAS TREATMENT
ANSI/ ASME N509-1980	Nuclear Power Plant Air Cleaning Units and Components
ANSI/ ASME N510-1980	Testing of Air-Cleaning Systems.
	NUCLEAR VALVES
N278.1-1975	Self-Operated and Power Operated Safety-Related Valves Functional Specification Standard.
N278.2-1983	Functional Qualification Requirements for Power Operated Active Valve Assemblies for Nuclear Power Plants.

Serials Published by IEEE

The Institute of Electrical and Electronic Engineers, with headquarters at 345 East 47th Street, New York, NY 10017, has been holding annual nuclear symposia since 1954 and publishing the transactions, first entitled *IRE Transactions on Nuclear Science*. From 1962 onward, the transactions have appeared under the title *IEEE Transactions on Nuclear Science*, and they report the latest developments in nuclear engineering as related to instrumentation, control systems and fusion.

Nuclear-related publications of IEEE are:

Title	Volume & Period
• *Proceedings*	51–, 1963–
• *Spectrum*	1–, 1964–
• *Transactions on Computers*	17–, 1968–
• *Control Instrumentation*	12–, 1965–
• *Transactions on Nuclear Science*	1–, 1955–
• *Transactions on Power Apparatus and Systems*	82–, 1963–
• *Transactions on Reliability*	17–, 1968–

Development of Standards by IEEE

The design of electronic and control devices is covered by the standards issued by the Institute of Electrical and Electronic Engineers. So far 55 nuclear and radiation instrumentation standards have been issued by the Institute. This number is expected to rise significantly in the near future, in view of the heavy involvement by members of the Institute in development work in the area of nuclear fusion. To underscore their growing importance, it is worth observing that ten of the IEEE nuclear standards have been invoked by the Nuclear Regulatory Commission in several of its Regulatory Guides.

Nuclear standards activities of IEEE are the responsibility of the Nuclear Power Engineering Committee, which in turn acts through subcommittees on: general plant criteria; qualification of electrical equipment; auxiliary power; quality assurance; reliability; operations, surveillance and testing; safety related systems; human factors and control facilities.

The nuclear standards of IEEE fall into the following categories:

- Controls
- Instrumentation
- Electrical systems for nuclear power generating stations
- Nuclear instruments

For ease of use, IEEE in 1981 published the full text of its standards[1] in a two-volume collection. Titles of IEEE standards currently in force are listed below. Standards in the process of approval, or under revision, have not been included.

REFERENCES

1. "Nuclear IEEE Standards," SHO8250, 2 vols., Institute of Electrical and Electronic Engineers, New York (June 1981).

Listing of Nuclear Standards Issued by IEEE

CONTROLS AND INSTRUMENTATION	
IEEE 279-1970(78)	Criteria for Protection Systems for Nuclear Power Generating Stations.
IEEE 336-1980	Installation Inspection and Testing Requirements for Class 1E Instrumentation and Electric Equipment at Nuclear Power Generating Stations.
IEEE 344-1975(80)	Seismic Qualification of Class 1E Electric Equipment for Nuclear Power Generating Stations.
IEEE 381-1977	Criteria for Type Tests of Class 1E Modules Used in Nuclear Power Generating Stations.
IEEE 383-1974(82)	Qualification of Class 1E Cables and Field Splices for Nuclear Power Generating Stations.
IEEE 485-1978	Sizing Large Lead Storage Batteries for Generating Stations and Substations.
IEEE/ ANS-7-4.3.2-1982	Application of Criteria for Programmable Digital Computer Systems in Safety Systems of Nuclear Power Generating Stations.
IEEE 352-1975(80)	Guide for General Principles of Reliability Analysis of Nuclear Power Generating Station Protection Systems.
IEEE 334-1974(82)	Type Tests of Continuous Duty Class 1E Motors for Nuclear Power Generating Stations.
IEEE 308-1980	Criteria for Class 1E Power Systems for Nuclear Power Generating Stations.
IEEE 535-1979	Qualification of Class 1E Lead Storage Batteries for Nuclear Power Generating Stations.
IEEE 622-1979	Recommended Practice for the Design and Installation of Electric Pipe Heating Systems for Nuclear Power Generating Stations.
IEEE 650-1979	Qualification of Class 1E Battery Chargers and Inverters for Nuclear Power Generating Stations.

NUCLEAR INSTRUMENTS	
N13.4-1971(77)	Specification of Portable X- or Gamma-Radiation Survey Meters.
N42.4-1971(78)	High Voltage Connectors for Nuclear Instruments.
N42.5-1965(77)	Base for GM Counter Tubes.
N42.6-1980	Interrelationship of Quartz-Fiber Electrometer Type Dosimeters and Companion Dosimeter Chargers.

Listing of Nuclear Standards Issued by IEEE (Continued)

N42.12-1980	Calibration and Usage of Sodium Iodide Detector Systems.
N42.13-1978	Calibration and Usage of "Dose Calibrator" Ionization Chambers for the Assay of Radionuclides.
N42.14-1978	Calibration and Usage of Germanium Detectors for Measurement of Gamma-Ray Emission of Radionuclides.
N42.15-1980	Performance Verification of Liquid Scintillation Counting Systems.
N42.18-1974(80)	Specification and Performance of On-Site Instrumentation for Continuously Monitoring Radioactivity in Effluents.
N317-1980	Performance Criteria for Instrumentation Used for In Plant Pu Monitoring.
N320-1979	Performance Specifications for Reactor Emergency Radiological Monitoring Instrumentation.
N322-1977	Inspection and Test Specifications for Direct and Indirect Reading Pocket Dosimeters.
N323-1978	Radiation Protection Instrumentation Test and Calibration.
N544-1968(79)	Signal Connectors for Nuclear Instruments.
IEEE 300-1982	Test Procedures for Semiconductor Charged-Particle Detectors.
IEEE 309-1970(81)	Test Procedures for Geiger-Muller Counters.
IEEE 325-1971(77)	Test Procedures for Germanium Gamma-Ray Detectors.
IEEE 398-1972(77)	Test Procedures for Photomultipliers for Scintillation Counting and Glossary for Scintillation Counting Field.
IEEE 583-1982	Modular Instrumentation and Digital Interface System (CAMAC).*
IEEE 595-1982	Serial Highway Interface System (CAMAC).
IEEE 596-1982	Parallel Highway Interface System (CAMAC).
IEEE 675-1982	Multiple Controllers in a CAMAC Crate.
IEEE 680-1978	Germanium Semiconductor Detector Gamma-Ray Efficiency Determination Using a Standard Marinelli (Reentrant) Beaker Geometry.
IEEE 683-1976(81)	Block Transfers in CAMAC Systems.
IEEE 726-1982	Real-Time BASIC for CAMAC.
IEEE 758-1979(81)	Subroutines for CAMAC.
IEEE 759-1982	Test Procedures for Semiconductor X-Ray Emergency Spectrometers.

* CAMAC = Computer Automated Measurement and Control.

Development of Standards by ASTM

The American Society for Testing and Materials, 1916 Race Street, Philadelphia, PA 19103, has been very active in the nuclear standards area. Nuclear standards activities of ASTM are concentrated in two of its committees: Nuclear Fuel Cycle (C-26) and Nuclear Technology and Applications (E-10). The standards developed by these committees deal with properties of materials used in the nuclear industry and with testing procedures developed to determine these properties.

In 1982 ASTM discontinued its prior practice, and ceased to submit to the American Standards Institute its new, revised or reapproved standards. This meant that they would no longer be approved as American National Standards. Time will show if this new stance will become a permanent policy of ASTM.

Some of the standards issued by the American Society for Testing and Materials for nuclear applications fall under the following headings: stainless steels and alloys, reactive and refractory metals, carbon and graphite products, concrete and concrete aggregates. The largest group of nuclear ASTM standards is that for the nuclear fuel cycle. Many other materials and components are covered, making the ASTM nuclear standards more numerous than those of any other society. ASTM maintains nuclear standards in the following areas:

- Nuclear fuel cycle
- Nuclear technology and applications
- Steel, stainless steel and related alloys
- Reactive and refractory metals and alloys
- Manufactured carbon and graphite products
- Concrete and concrete aggregates
- Protective coatings
- Water
- Temperature measurement
- Criteria for the evaluation of test and inspection agencies

Titles of ASTM nuclear standards currently in use are listed below. Standards under revision or in the process of approval were not included, nor were the standards for water chemistry and protective coatings. The latter two groups fall more into the area of radiochemistry than nuclear engineering proper and for this reason are considered outside the scope of this guide.

Individual ASTM standards sell at a uniform price of $8.00 per copy. The 107 standards developed by ASTM Committee C-26 on the nuclear fuel cycle have been published as section 12.01 of the ASTM book of standards[(1)], and those of Committee E-10 on nuclear technology and applications are in section 12.02, containing nuclear, solar and geothermal energy standards[(2)]. Summaries of radiation effects have also been issued in book form.[(3)–(4)]

REFERENCES

1. "1983 Annual Book of ASTM Standards, Nuclear Energy (I)," Vol. 12.01, ASTM, Philadelphia (Sept. 1983).
2. "1983 Annual Book of ASTM Standards, Nuclear (II), Solar and Geothermal Energy," Vol. 12.02, ASTM, Philadelphia (Sept. 1983).

3. L. E. Steele, ed., "Radiation Embrittlement and Surveillance of Nuclear Reactor Pressure Vessels: An International Study," STP 819, ASTM, Philadelphia (1983).
4. D. Franklin, ed., "Zirconium in the Nuclear Industry," STP 754, ASTM, Philadelphia (1982).

Listing of Nuclear Standards Issued by ASTM

STAINLESS STEEL AND RELATED ALLOYS	
A614-79	Special Requirements for Bolting Material for Nuclear and Other Special Applications.
A652-80	Special Requirements for Wrought Steel Welding Fittings for Nuclear and Other Special Applications.
A654-79	Special Requirements for Forgings and Bars for Nuclear and Other Special Applications.
A655-77	Special Requirements for Pipe and Tubing for Nuclear and Other Special Applications.
A771-80	Austenitic Stainless Steel Tubing for Breeder Reactor Core Components.
REACTIVE AND REFRACTORY METALS AND ALLOYS	
B349-80	Specification for Zirconium Sponge and Other Forms of Virgin Metal for Nuclear Applications.
B350-80	Specification for Zirconium and Zirconium Alloy Ingots for Nuclear Application.
B351-79	Specification for Hotrolled and Coldfinished Zirconium and Zirconium Alloy Bars, Rod, and Wire for Nuclear Application.
B352-79	Specification for Zirconium and Zirconium Alloy Sheet, Strip, and Plate for Nuclear Application.
B353-77a	Specification for Wrought Zirconium and Zirconium Alloy Seamless and Welded Tubes for Nuclear Service.
B600-74	Recommended Practice for Descaling and Cleaning Titanium and Titanium Alloy Surfaces.
MANUFACTURED CARBON AND GRAPHITE PRODUCTS	
C625-72	Reporting Irradiation Results on Graphite.
C626-71(77)	Estimating the Thermal Neutron Absorption Cross Section of Nuclear Graphite.
C709-77	Definitions of Terms Relating to Manufactured Carbon and Graphite.
C747-74(79)	Method for Sonic Determination of Moduli of Elasticity and Fundamental Frequencies of Carbon and Graphite Materials.
C769-80	Sonic Velocity in Manufactured Carbon and Graphite Materials for Use in Obtaining an Approximate Young's Modulus.
C781-77	Recommended Practice for Testing Graphite and Boronated Graphite for HTGR's.

Listing of Nuclear Standards Issued by ASTM (Continued)

C783-74(80)	Recommended Practice for Core Sampling of Graphite Electrodes.
C816-77	Method of Test for Total Boron in Graphite.
C859-78	Definition of Terms Relating to Nuclear Materials.
	NEUTRON ABSORBER MATERIALS
C708-72a(77)	Specification for Nuclear Grade Beryllium Oxide Powder.
C750-80	Specification for Nuclear Grade Boron Carbide Powder.
C751-75(81)	Specification for Nuclear Grade Boron Carbide Pellets.
C752-79	Specification for Nuclear Grade Silver–Indium–Cadmium Alloy.
C784-76	Specification for Nuclear Grade Aluminum Oxide–Boron Carbide Composite Pellets.
	NUCLEAR FUELS
C753-78	Specification for Nuclear Grade Sinterable Uranium Dioxide Powder.
C757-79	Standard Specification for Nuclear Grade Sinterable Plutonium Dioxide Powder.
C785-76	Specification for Nuclear Grade Aluminum Oxide Pellets.
C776-79	Specifications for Sintered Uranium Dioxide Pellets.
C787-76(80)	Specification for Uranium Hexafluoride for Enrichment.
C833-76(80)	Specification for Sintered (Uranium–Plutonium) Pellets.
C922-79	Sintered Gadolinium–Uranium Dioxide Pellets.
C934-80	Design and Quality Assurance Practices for Nuclear Fuel Rods.
C967-81	Specifications for Uranium Ore Concentrates.
	CONCRETE AND CONCRETE AGGREGATES
C637-73(79)	Specification for Aggregates for Radiation Shielding Concrete.
C638-73(79)	Descriptive Nomenclature of Constituents of Aggregates for Radiation Shielding Concrete.
E669-79	Master Matrix for Nuclear Fuel Cycle Standards.
E846	Matrix for Light Water Reactor Spent Fuel Transportation.
E235-76	Thermocouples Sheathed Type K for Nuclear or High Reliability Application.
	TEST METHODS
C696-80	Method for Chemical, Mass Spectrometric and Spectrochemical Analysis of Nuclear-Grade Uranium Dioxide Powder and Pellets.
C697-80a	Method for Chemical, Mass Spectrometric and Spectrochemical Analysis of Nuclear-Grade Plutonium Dioxide Powders and Pellets.
C698-80a	Method for Chemical, Mass Spectrometric and Spectrochemical Analysis of Nuclear-Grade Mixed Oxides $(U,Pu)O_2$.
C699-79	Methods for Chemical, Mass Spectrometric and Spectrochemical Analysis of, and Physical Tests on Beryllium Oxide Powder.
C758-79	Method for Chemical, Mass Spectrometric, Spectrochemical Nuclear and Radiochemical Analysis of Nuclear Grade Plutonium Metal.

Listing of Nuclear Standards Issued by ASTM (Continued)

C759-79	Method for Chemical, Mass Spectrometric, Spectrochemical Nuclear and Radiochemical Analysis of Nuclear Grade Plutonium Nitrate Solution.
C760-79	Method for Chemical and Spectrochemical Analysis of Nuclear Grade Silver–Cadmium–Indium Alloys.
C761-77	Method for Chemical, Mass Spectrometric and Spectrochemical Analysis of Uranium Hexafluoride.
C791-79	Methods of Test for Chemical, Mass Spectrometric, and Spectrochemical Analysis of Nuclear-Grade Boron Carbide.
C799-78	Methods of Test for Chemical, Mass Spectrometric, Spectrochemical, Nuclear, and Radiochemical Analysis of Nuclear Grade Uranyl Nitrate Solutions.
C809-80	Chemical, Mass Spectrometric and Spectrochemical Analysis of Nuclear Grade Aluminum Oxide and Aluminum Oxide Boron Carbide Pellets.
C852-77	Design Criteria for Plutonium Glove Boxes.
C853-79	Method for Nondestructive Assay of Special Nuclear Materials Contained in Scrap and Waste.
C888-78	Specification for Nuclear Grade Gadolinium Oxide (Gd_2O_3) Powder.
C889-78	Methods for Chemical and Spectrochemical Analysis of Nuclear Grade Gadolinium Oxide (Gd_2O_3) Powder.
NUCLEAR TECHNOLOGY AND APPLICATIONS	
E217-70(78)	Method of Test for Uranium by Controlled-Potential Coulometry.
E219-80	Method of Test for Atom Percent Fission in Uranium Fuel (Radio-Chemical Method).
E244-80	Method of Test for Atom Percent Fission in Uranium and Plutonium Fuel (Mass Spectrometric Method).
E267-78	Method of Test for Uranium and Plutonium Concentrations and Isotopic Abundances.
E318-69(80)	Method for Colorimetric Determination of Uranium in Aqueous Solutions.
E320-79	Methods for Radiochemical Determination of Cesium-137 in Nuclear Fuel Solutions.
E321-79	Method of Test for Atom Percent Fission in Uranium and Plutonium Fuel (Neodymium-148 Method).
E495-78	Spectrophotometric Determination of Fission Zirconium in Irradiated Nuclear Fuels.
E692-79	High Resolution Gamma-Ray Spectral Analysis for Cesium-137 in Irradiated Nuclear Fuels.
BEHAVIOR OF METALLIC MATERIALS IN NUCLEAR SYSTEMS	
E184-79	Effects of High-Energy Neutron Radiation on the Mechanical Properties of Metallic Materials.
E185-79	Conducting Surveillance Tests for Light-Water Cooled Nuclear Power Reactor Vessels.
E509-74(80)	In-Service Annealing of Water-Cooled Nuclear Reactor Vessels.

Listing of Nuclear Standards Issued by ASTM (Continued)

E531-76(80)	Surveillance Testing of High-Temperature Nuclear Component Materials.

NEUTRON FLUX DETERMINATION

E261-77	Neutron Flux, Fluence, and Spectra by Radioactivation Techniques.
E262-77	Thermal Neutron Flux by Radioactivation Techniques.
E263-77	Fast-Neutron Flux Density by Radioactivation of Iron.
E264-77	Fast Neutron Flux Density by Radioactivation of Nickel.
E265-77	Fast-Neutron Flux Density by Radioactivation of Sulfur.
E266-77	Fast-Neutron Flux Density by Radioactivation of Aluminum.
E385-80	Oxygen Content Using a 14-MeV Neutron Activation and Direct Counting Technique.
E393-79	Fast-Neutron Flux Density by Analysis for Ba-140 Produced by U-238 Fission.
E418-73(79)	Fast-Neutron Flux Measurements by Trace-Etch Technique.
E419-73(79)	Selection of Neutron Activation Detector Materials.
E481-78	Neutron-Flux Density by Radioactivation of Cobalt and Silver.
E482-76	Guide for Application of Neutron Transport Methods for Reactor Vessel Surveillance.
E496-79	Neutron-Flux Density and Average Energy from $^3H(d,N)^4He$ Neutron Generators by Radioactivation Techniques.
E522-78	Calibration of Germanium Detectors for Measurement of Gamma-Ray Emission Rates of Radionuclides.
E523-76	Determining Fast Neutron Flux Density by Radioactivation of Copper.
E525-74	Reporting Neutron Dosimetry Results on Nuclear Graphite.
E560-77	Extrapolating Reactor Vessel Surveillance Dosimetry Results.
E693-79	Characterizing Neutron Exposures in Ferritic Steel in Terms of Displacements per Atom.
E704-79	Fast-Neutron Flux Density by Radioactivation of Uranium-238.
E705-79	Fast-Neutron Flux Density by Radioactivation of Neptunium-237.
E706-79	Master Matrix of Procedures for Light-Water Reactor Pressure Vessel Surveillance Standard.
E854	Application and Analysis of Solid State Track Recorder (SSTR) Monitors for Reactor Vessel Surveillance.

RADIATION DOSIMETRY

E665-78	Determining Absorbed Dose Versus Depth in Materials Exposed to the X-Ray Output of Flash X-Ray Machines.
E666-78	Calculation of Absorbed Dose from Gamma or X-Radiation.
E668-78	Thermoluminescence-Dosimetry (TLD) Systems for Determining Absorbed Dose in Radiation-Hardness Testing of Electronic Devices.
E720-80	Selection of a Set of Neutron Activation Foils for Determining Neutron Spectra Used in Radiation-Hardness Testing of Electronics.
E721-80	Neutron Activation Foils for Neutron Energy Spectra Radiation Hardness Testing of Electronics.
E722-80	Characterizing Neutron Energy Fluence Spectra in Terms of An Equivalent Mono-Energetic Neutron Fluence for Radiation Hardness Testing of Electronics.

Listing of Nuclear Standards Issued by ASTM (Continued)

E763-80	Calculation of Absorbed Dose in Materials from Irradiation by Application of Threshold-Foil Measurement Data. E845 Calibration of Dosimeters Against an Adiabatic Calorimeter for Use in Flash X-Ray Fields.
E583-77	Systematizing the Development of (ASTM) Voluntary Consensus Standards for the Solution of Nuclear and Other Complex Problems.
E548-79	Generic Criteria for Evaluation of Testing and/or Inspection Agencies.
E764-80	Guide for Identifying Test Methods for Breeder Reactor Core Materials.

Nuclear Standards Developed by Other Societies

Several other professional societies are active in the development of nuclear standards, either as their main task or as part of their regular business. In the first of these categories are the Health Physics Society (HPS) and the Institute of Nuclear Materials Management (INMM).

Among societies that have only one or two sections engaged in the development of nuclear standards are: the Instrument Society of America (ISA), American Concrete Institute (ACI), American Insurance Association (AIA), American Institute of Chemical Engineers (AIChE), American Institute of Steel Construction (AISC), American Society of Civil Engineers (ASCE), National Fire Protection Association (NFPA) and National Bureau of Standards (NBS). Not all of these, however, have produced standards that have been already approved, or are in general use by the industry.

Neither proposed standards nor standards under revision have been included in the current lists of titles presented below. It should be pointed out, however, that some proposed standards have already been quoted in several proposed regulatory guides. In spite of this, their inclusion here cannot be justified until they have received final approval.

Listing of Nuclear Standards Issued by NBS

EQUIPMENT FOR NONMEDICAL RADIATION APPLICATIONS	
N43.1-1978	Radiological Safety in the Design and Operation of Particle Accelerators (NBS Handbook 107).
N43.2-1977	Radiation Safety for X-Ray Diffraction and Fluorescence Analysis Equipment (NBS Handbook 111).
N432-1980	Radiological Safety for the Design and Construction of Apparatus for Gamma Radiography (NBS Handbook 136).
N4331-1977	Sealed Source Irradiators—Category 1 (NBS Handbook 127).
N538-1979	Classification of Industrial Ionizing Radiation Gauging Devices (NBS Handbook 129).
N542-1977	Classification of Sealed Radioactive Sources (NBS Handbook 126).

Listing of Nuclear Standards Issued by HPS

RADIATION PROTECTION	
N13.1-1969(82)	Guide to Sampling Airborne Radioactive Materials in Nuclear Facilities.
N13.2-1969(82)	Guide for Administrative Practices in Radiation Monitoring.
N13.3-1969(81)	Dosimetry for Criticality Accidents.
N13.5-1972(82)	Performance Specifications for Direct Reading and Indirect Reading Pocket Dosimeters for X- and Gamma-Radiation.
N13.6-1966(82)	Practice for Occupational Radiation Exposure Records System.
N13.7-1981	Criteria for Film Dosimeter Performance.
N13.8-1973(82)	Radiation Protection in Uranium Mines Operation.
N13.14-1981	Internal Dosimetry Techniques for Tritium.
N13.15-1981	Performance of Personnel Thermoluminescence Dosimetry Systems.
N13.27-1981	Performance Specifications for Pocket-Sized Alarming Dosimeters/ Ratemeters.
N343-1978	Internal Dosimetry Techniques for Fission and Activation Products.
N545-1975(82)	Environmental Application of Thermoluminescent Dosimetry.

Listing of Nuclear Standards Issued by INMM

TRANSPORTATION OF FISSILE AND RADIOACTIVE MATERIALS	
N14.1-1975 (82)	Packaging of Uranium Hexafluoride for Transport.
N14.6-1978	Special Lifting Devices for Shipping Containers Weighing 10,000 Pounds (4,500 kg) or More for Nuclear Materials.
N679-1976	Guide for Writing Operating Manuals for Radioactive Materials Packaging.
N15.15-74(81)	Assessment of the Assumption of Normality (Employing Individual Observed Values).
N15.18-1975	Mass Calibration Techniques for Nuclear Material Control.
N15.23-1979	Guide to the Nondestructive Assay of ^{235}U Content of Unpoisoned Low-Enrichment Uranium Fuel Rods.
N15.35-1982	Guide to Preparing Calibration Material for Nondestructive Assay by Counting Passive Gamma Rays.
N15.36-1983	Nondestructive Assay Measurement Control and Assurance.
N15.37-1981	Guide to the Automation of Nondestructive Assay Systems for Nuclear Material Control.
N15.38-1982	A Generic Standard for Auditing Nuclear Materials Safeguards Systems.
N15.40-1981	Definition of Terms and Symbols Associated with the Physical Protection of Nuclear Materials and Facilities.

Listing of Nuclear Standards Issued by ISA

S 18.1-1979	Annunciator Sequences and Specifications.
S 67.01-1979	Standard Transducer and Transmitter Installation for Nuclear Safety Applications.

Listing of Nuclear Standards Issued by AIChE

NUCLEAR REACTOR FUEL CYCLE	
N46.1-1981	Criteria for Determining Safety-Related Features of Nuclear Fuel Cycle Facilities.
N46.2-1978	QA Program Requirements for Post Reactor Nuclear Fuel Cycle Facilities.
N101.1-1972	Efficiency Testing of Air Cleaning Systems Containing Devices for Removal of Particles.
N101.3-1979	Guide to Principal Design Criteria, Nuclear Fuel Reprocessing Facilities.
N287-1976	Criteria for Siting, Design, and Operation of Plants for the Manufacture of Mixed Oxide (UPu) Fuels.
N290-1979	Design, Construction and Operation of Ventilation Systems for Mixed Oxide(UO_2-PuO_2) Fuel Fabrication Plants.
N300-1978	Design Criteria for Decommissioning of Nuclear Fuel Reprocessing Facilities.
N303-1978	Guide for Control of Gasborne Radioactive Materials at Fuel Reprocessing Facilities.
N304-1978	Fire Protection for Fuel Reprocessing Facilities.
N305-1975(81)	Design Objectives for Highly Radioactive Solid Material Handling and Storage Facilities in a Reprocessing Plant.

Listing of Nuclear Standards Issued by NFPA

NFPA-801-1980	Fire Protection Practice for Facilities Handling Radioactive Materials.

9. FEDERAL REGULATIONS, GUIDES AND SERIALS

9.1 REGULATORY ACTIVITIES OF FEDERAL GOVERNMENT

In addition to the large body of standards connected with nuclear engineering, there exists an expanding collection of Regulatory Guides and Federal Regulations, issued by the Nuclear Regulatory Commission. The Federal Regulations have their base in the legislation enacted by Congress, and are enforceable under law; i.e., noncompliance exposes the defaulting persons or organizations to prosecution through the courts. The Guides, on the other hand, do not have the force of law. Nevertheless, they carry behind them enough weight of the regulatory body to command attention.

Code of Federal Regulations

The Federal Regulations directly concerned with the construction and operation of nuclear reactors are listed in the Code of Federal Regulations (CFR), Title 10, Energy. The regulations are usually quite detailed and embody substantial technical knowledge in their formulation. They can be quoted in design as a justification for specific design decisions.

Topics of Title 10 of Code of Federal Regulations

0. Conduct of Employees
1. Statement of Organization and General Information
2. Rules of Practice for Domestic Licensing Proceedings
4. Nondiscrimination in Federally Assisted Commission Programs
7. Advisory Committees
8. Interpretations
9. Public Records
10. Criteria and Procedures for Determining Eligibility for Access to Restricted Data or National Security Information or an Employment Clearance

11. Criteria and Procedures for Determining Eligibility for Access to or Control Over Unclassified Special Nuclear Material
14. Administrative Claims Under Federal Tort Claims Act
15. Debt Collection Procedures
19. Notices, Instructions, and Reports to Workers; Inspections
20. Standards of Protection Against Radiation
21. Reporting of Defects and Noncompliance
25. Access Authorization for Licensee Personnel
30. Rules of General Applicability to Domestic Licensing of Byproduct Material
31. General Domestic Licenses for Byproduct Material
32. Specific Domestic Licenses to Manufacture or Transfer Certain Items Containing Byproduct Material
33. Specific Domestic Licenses of Broad Scope for Byproduct Material
34. Licenses for Radiography and Radiation Safety Requirements for Radiographic Operations
35. Human Uses of Byproduct Material
40. Domestic Licensing of Source Material
50. Domestic Licensing of Production and Utilization Facilities
51. Licensing and Regulatory Policy and Procedures for Environmental Protection
55. Operators' Licenses
60. Disposal of High Level Radioactive Wastes in Geologic Repositories
61. Licensing Requirements for Land Disposal of Radioactive Wastes
70. Domestic Licensing of Special Nuclear Material
71. Packaging and Transportation of Radioactive Material
72. Licensing Requirements for the Storage of Spent Fuel in an Independent Spent Fuel Storage Installation (ISFSI)
73. Physical Protection of Plants and Materials
75. Safeguards on Nuclear Material—Implementation of US/IAEA Agreement
81. Standard Specifications for the Granting of Patent Licenses
95. Security Facility Approval and Safeguarding of National Security Information and Restricted Data
100. Reactor Site Criteria
110. Export and Import of Nuclear Equipment and Materials
140. Financial Protection Requirements and Indemnity Agreements
150. Exemptions and Continued Regulatory Authority in Agreement States and in offshore Waters Under Section 274
160. Trespassing on Commission Property
170. Fees for Facilities and Materials Licenses and Other Regulatory Services under the Atomic Energy Act of 1954, as Amended

The parts of Title 10 on which the nuclear industry relies heavily are: Part 20, which deals with the standards for protection against radiation; Part 50, treating the domestic licensing of production and utilization facilities; Part 55, concerned with operator's licenses; and Part 100, which deals with reactor site criteria.

Nuclear Regulatory Guides

An interrelationship exists between the Federal Regulatory Guides and the standards and codes developed by the various engineering societies, insofar as some of these regulatory documents invoke the existing and generally accepted standards and codes. This includes the Boiler and Pressure Vessel Code of ASME, and some of the standards of IEEE, ASTM and ACI. Altogether, over 70 Regulatory Guides quote various engineering standards.

Whenever engineering performance and test standards are quoted in the Regulatory Guides, reference is made either to a complete code or relevant sections of it. This is particularly true of the Boiler and Pressure Vessel Code of ASME. Sections III, V, VIII, IX, X and XI of that code are quoted in 19 different Guides.

The Regulatory Guides are under continuous surveillance, and those judged outdated are revised every few years. Up to three revisions from the date of issue are not unusual. Of the 310 Federal Guides now in force, 150 have already been revised at least once. As for the guides incorporating Section III of the Boiler and Pressure Vessel Code of ASME, the revisions and updates are carried out annually, 19 of them so far. This is indicative of the importance and the central position held in the nuclear technology by reactor pressure vessels, and the attention given to maintaining their safe and secure operation. From time to time some Regulatory Guides are withdrawn from circulation, and the public is duly notified of the fact by an announcement placed in the Federal Register.

The Regulatory Guide series is published in ten major divisions. Each Guide is issued first in a draft form for comment, with the intent of involving the public in the early stages of the development of a regulatory position. When eventually issued in their final form, they represent an official NRC staff position on a specific licensing issue. For this reason, the Guides are diligently followed by the applicant for a license to assure expeditious treatment of the license application itself. This body of information is intended to provide guidance for the reactor designers and operators, as well as for the ancillary organizations of the nuclear industry. Material contained in the guides is arranged in divisions listed in Table 1.

Table 1. Divisions of Nuclear Regulatory Guides

Division	# of Guides	Subscription
1. Power Reactors	(150 guides)	$100.00
2. Research and Test Reactors	(6 guides)	$14.00
3. Fuels and Materials Facilities	(53 guides)	$60.00
4. Environment and Siting	(17 guides)	$28.00
5. Materials and Plant Protection	(63 guides)	$55.00
6. Products	(8 guides)	$14.00
7. Transportation	(10 guides)	$14.00
8. Occupational Health	(29 guides)	$35.00
9. Antitrust and Financial Review	(4 guides)	$13.00
10. General	(9 guides)	$37.00

As regards the actual content and form, the Guides fall into several well-defined categories, each designed to address a specific issue that might arise in any of the above ten divisions, as follows:

Quality Assurance. The Guide specifies the standards of workmanship, or material quality, that must be satisfied before the component is deemed acceptable for nuclear service.

Qualification Testing. The Guide describes the tests that must be carried out before quality or performance is deemed acceptable for nuclear service.

Assumptions for Evaluations. The Guide states formally the assumptions that are considered reasonable for the assessment of radiological consequences of failure of a particular piece of equipment.

Design Bases. The Guide describes the parameters for natural calamities, usually termed "acts of God," such as floods, tornados and earthquakes, based on data from past experience.

Emergency Planning. The Guide specifies the conditions to be satisfied in planning for emergency situations in nuclear reactor operations and fuel cycle facilities.

Standard Formats. The Guide outlines the format and content expected of documentation submitted to the NRC in support of applications for license to construct or operate nuclear facilities.

Reporting Procedures. The Guide details the formats and frequencies of reporting of activities or occurrences in plant operation of interest to the Commission.

Periodic Testing. The Guide recommends the frequency and the mode of testing for emergency backup equipment, normally remaining in the standby condition.

Design Limits. The Guide outlines the design considerations, design limits and criteria to be used in the preparation of plans and procedures in the nuclear industry.

Protection Guidelines. The Guide suggests methods of protecting the integrity of plants and equipment from consequences of accidents or willful sabotage actions.

Estimating Methods. The Guide states the approach that may be taken by the licensee in preparing the estimates of magnitude of a particular physical or chemical process.

Calculating Methods. The Guide describes the methods of computation available for designing a process or equipment to meet the regulatory criteria of performance or protection from operating hazards.

The usefulness of the material contained in the guides has proved itself many times in the past. By addressing the content of the Guides to the issues of common concern to the various industrial and utility companies operating in the nuclear field, the Nuclear Regulatory Commission has done much to place the analysis, planning and reporting activities of these organizations on a common footing. It must be observed, however, that there are certain risks involved in the specification of conditions of accidental occurrences for the purpose of analysis. The recent analysis of the consequences of the accident in the Three Mile Island reactor indicated the need for the revision of some of the assumptions used in the Guides. Comparison of the predicted degree of release of the gaseous fission product inventory contained in the fuel with the actual release in consequence of core damage revealed a sizable overestimate of the release fraction used in the assumptions recommended by the NRC for safety analyses. A reexamination of the assumptions is now in progress.

The number, title and year of latest revision of the currently used guides are listed on the following pages. The Guides are numbered sequentially within each division. The missing Guide numbers in the sequences correspond to Guides withdrawn by the Commission, with the withdrawal noted in each case in the Federal Register. The Guides themselves may be inspected at the Public Document Room of NRC at 1717 H Street NW, Washington, DC 20555, or may be purchased from the Government Printing Office. Subscription service consists of the basic manual and supplementary material for an indeterminate period at prices listed in the last column of Table 1 above. Supplements only are available separately at prices varying from $1.75 to $4.00.

Nuclear Regulatory Guides for Power Reactors

GUIDE/YR	REGULATORY GUIDES, POWER REACTORS
1.1 1970	Net Positive Suction Head for Emergency Core Cooling and Containment Heat Removal System Pumps
1.2 1970	Thermal Shock to Reactor Pressure Vessels
1.3 1974	Assumptions Used for Evaluating the Potential Radiological Consequences of a Loss-of-Coolant Accident for Boiling Water Reactors
1.4 1974	Assumptions Used for Evaluating the Potential Radiological Consequences of a Loss-of-Coolant Accident for Pressurized Water Reactors
1.5 1971	Assumptions Used for Evaluating the Potential Radiological Consequences of a Steam Line Break Accident for Boiling Water Reactors
1.6 1971	Independence Between Redundant Standby (Onsite) Power Sources and Between Their Distribution Systems
1.7 1978	Control of Combustible Gas Concentrations in Containment Following a Loss-of-Coolant Accident
1.8 1977	Personnel Selection and Training
1.9 1979	Selection, Design, and Qualification of Diesel-Generator Units Used as Standby (Onsite) Electric Power Systems at Nuclear Power Plants
1.11 1971/72	Instrument Lines Penetrating Primary Reactor Containment and Supplement to Safety Guide 1.11: Backfitting Considerations
1.12 1974	Instrumentation for Earthquakes
1.13 1975	Spent Fuel Storage Facility Design Basis
1.14 1975	Reactor Coolant Pump Flywheel Integrity
1.16 1975	Reporting of Operating Information, Appendix A, Technical Specification
1.17 1973	Protection of Nuclear Power Plants Against Industrial Sabotage
1.20 1976	Comprehensive Vibration Assessment Program for Reactor Internals During Preoperational and Initial Startup Testing
1.21 1974	Measuring, Evaluating, and Reporting Radioactivity in Solid Wastes and Release of Radioactivity in Liquid and Gaseous Effluents from Light-Water-Cooled Nuclear Power Plants
1.22 1972	Periodic Testing of Protection System Actuation Functions
1.23 1972	Onsite Meteorological Programs
1.24 1972	Assumptions Used For Evaluating the Potential Radiological Consequences of a Pressurized Water Reactor Gas Storage Tank Failure

Nuclear Regulatory Guides for Power Reactors (Continued)

1.25 1972	Assumptions Used for Evaluating the Potential Radiological Consequences of a Fuel Handling Accident in the Fuel Handling and Storage Facility for Boiling and Pressurized Water Reactors
1.26 1976	Quality Group Classifications and Standards for Water-, Steam-, and Radioactive-Waste-Containing Components of Nuclear Power Plants
1.27 1976	Ultimate Heat Sink for Nuclear Power Plants
1.28 1979	Quality Assurance Program Requirements (Design and Construction)
1.29 1978	Seismic Design Classification
1.30 1972	Quality Assurance Requirements for the Installation, Inspection and Testing of Instrumentation and Electric Equipment
1.31 1978	Control of Ferrite Content in Stainless Steel Weld Metal
1.32 1977	Criteria for Safety-Related Electric Power Systems for Nuclear Power Plants
1.33 1978	Quality Assurance Program Requirements (Operation)
1.34 1972	Control of Electroslag Weld Properties
1.35 1976	Inservice Inspection of Ungrouted Tendons in Prestressed Concrete Containment Structures
1.36 1973	Nonmetallic Thermal Insulation for Austenitic Stainless Steel
1.37 1973	Quality Assurance Requirements for Cleaning of Fluid Systems and Associated Components of Water-Cooled Nuclear Power Plants
1.38 1977	Quality Assurance Requirements for Packaging, Shipping, Receiving, Storage, and Handling of Items for Water-Cooled Nuclear Power Plants
1.39 1977	Housekeeping Requirements for Water-Cooled Nuclear Power Plants
1.40 1973	Qualification Tests of Continuous-Duty Motors Installed Inside the Containment of Water-Cooled Nuclear Power Plants
1.41 1973	Preoperational Testing of Redundant On-Site Electric Power Systems to Verify Proper Load Group Assignments
1.43 1973	Control of Stainless Steel Weld Cladding of Low-Alloy Steel Components
1.44 1973	Control of the Use of Sensitized Stainless Steel
1.45 1973	Reactor Coolant Pressure Boundary Leakage Detection System
1.46 1973	Protection Against Pipe Whip Inside Containment
1.47 1973	Bypassed and Inoperative Status Indication for Nuclear Power Plant Safety Systems
1.48 1973	Design Limits and Loading Combinations for Seismic Category I Fluid Systems Components

Nuclear Regulatory Guides for Power Reactors (Continued)

1.49 1973	Power Levels of Nuclear Power Plants
1.50 1973	Control of Preheat Temperature for Welding of Low-Alloy Steel
1.52 1978	Design, Testing, and Maintenance Criteria for Post Accident Engineered-Safety-Feature Atmosphere Cleanup System Air Filtration and Adsorption Units of Light-Water-Cooled Nuclear Power Plants
1.53 1973	Application of the Single-Failure Criterion to Nuclear Power Plant Protection Systems
1.54 1973	Quality Assurance Requirements for Protective Coatings Applied to Water-Cooled Nuclear Power Plants
1.56 1978	Maintenance of Water Purity in Boiling Water Reactors
1.57 1973	Design Limits and Loading Combinations for Metal Primary Reactor Containment System Components
1.58 1980	Qualification of Nuclear Power Plant Inspection, Examination, and Testing Personnel
1.59 1977/80	Design Basis Floods for Nuclear Power Plants, and Errata to Regulatory Guide 1.59
1.60 1973	Design Response Spectra for Seismic Design of Nuclear Power Plants
1.61 1973	Damping Values for Seismic Design of Nuclear Power Plants
1.62 1973	Manual Initiation of Protective Actions
1.63 1978	Electric Penetration Assemblies in Containment Structures for Light-Water Cooled Nuclear Power Plants
1.64 1976	Quality Assurance Requirements for the Design of Nuclear Power Plants
1.65 1973	Materials and Inspections for Reactor Vessel Closure Studs
1.68 1978	Initial Test Programs for Water-Cooled Nuclear Power Plants
1.68.1 1977	Preoperational and Initial Startup Testing of Feedwater and Condensate Systems for Boiling Water Reactor Power Plants
1.68.2 1978	Initial Startup Test Program to Demonstrate Remote Shutdown Capability for Water-Cooled Nuclear Power Plants
1.68.3 1982	Preoperational Testing of Instrument and Control Air Systems
1.69 1973	Concrete Radiation Shields for Nuclear Power Plants
1.70 1978	Standard Format and Content of Safety Analysis Reports for Nuclear Power Plants
1.71 1973	Welder Qualification for Areas of Limited Accessibility
1.72 1978	Spray Pond Piping Made from Fiberglass-Reinforced Thermosetting Resin
1.73 1974	Qualification Tests of Electric Valve Operators Installed Inside the Containment of Nuclear Power Plants

Nuclear Regulatory Guides for Power Reactors (Continued)

1.74 1974	Quality Assurance Terms and Definitions
1.75 1978	Physical Independence of Electric Systems
1.76 1974	Design Basis Tornado for Nuclear Power Plants
1.77 1974	Assumptions Used for Evaluating a Control Rod Ejection Accident for Pressurized Water Reactors
1.78 1974	Assumptions for Evaluating the Habitability of a Nuclear Power Plant Control Room During a Postulated Hazardous Chemical Release
1.79 1975	Preoperational Testing of Emergency Core Cooling Systems for Pressurized Water Reactors
1.81 1975	Shared Emergency and Shutdown Electric Systems for Multi-Unit Nuclear Power Plants
1.82 1974	Sumps for Emergency Core Cooling and Containment Spray Systems
1.83 1975	Inservice Inspection of Pressurized Water Reactor Steam Generator Tubes
1.84 1983	Design and Fabrication Code Case Acceptability—ASME Section III, Division 1
1.85 1983	Materials Code Case Acceptability—ASME Section III, Division 1
1.86 1974	Termination of Operating Licenses for Nuclear Reactors
1.87 1975	Guidance for Construction of Class I Components in Elevated-Temperature Reactors (Supplement to ASME Section III Code Classes 1592, 1593, 1594, 1595 and 1596)
1.88 1976	Collection, Storage, and Maintenance of Nuclear Power Plant Quality Assurance Records
1.89 1974	Qualification of Class 1E Equipment for Nuclear Power Plants
1.90 1977	Inservice Inspection of Prestressed Concrete Containment Structures with Grouted Tendons
1.91 1978	Evaluation of Explosions Postulated to Occur on Transportation Routes Near Nuclear Power Plants
1.92 1976	Combining Modal Responses and Spatial Components in Seismic Response Analysis
1.93 1974	Availability of Electric Power Sources
1.94 1976	Quality Assurance Requirements for Installation, Inspection, and Testing of Structural Concrete and Structural Steel During the Construction Phase of Nuclear Power Plants
1.95 1977	Protection of Nuclear Power Plant Control Room Operators Against an Accidental Chlorine Release
1.96 1976	Design of Main Steam Isolation Valve Leakage Control Systems for Boiling Water Reactor Nuclear Power Plants
1.97 1983	Instrumentation for Light-Water-Cooled Nuclear Power Plants to Assess Plant and Environs Conditions During and Following an Accident

Nuclear Regulatory Guides for Power Reactors (Continued)

1.98 1976	Assumptions Used for Evaluating the Potential Radiological Consequences of Radioactive Offgas System Failure in a Boiling Water Reactor
1.99 1977	Effects of Residual Elements on Predicted Radiation Damage to Reactor Vessel Materials
1.100 1977	Seismic Qualification of Electric Equipment for Nuclear Power Plants
1.101 1981	Emergency Planning and Preparedness for Nuclear Power Reactors
1.102 1976	Flood Protection for Nuclear Power Plants
1.105 1976	Instrument Setpoints
1.106 1977	Thermal Overload Protection for Electric Motors on Motor-Operated Valves
1.107 1977	Qualifications for Cement Grouting for Prestressing Tendons in Containment Structures
1.108 1977	Periodic Testing of Diesel Generator Units Used as Onsite Electric Power Systems at Nuclear Power Plants, and Errata
1.109 1977	Calculation of Annual Doses to Man from Routine Releases of Reactor Effluents for the Purpose of Evaluating Compliance with 10 CFR Part 50, Appendix I
1.110 1976	Cost-Benefit Analysis for Radwaste Systems for Light-Water-Cooled Nuclear Power Reactors
1.111 1977	Methods for Estimating Atmospheric Transport and Dispersion of Gaseous Effluents in Routine Releases from Light-Water-Cooled Reactors
1.112 1977	Calculation of Releases of Radioactive Materials in Gaseous and Liquid Effluents from Light-Water-Cooled Power Reactors
1.113 1977	Estimating Aquatic Dispersion of Effluents from Accidental and Routine Reactor Releases for the Purpose of Implementing Appendix I
1.114 1976	Guidance on Being Operator at the Controls of a Nuclear Power Plant
1.115 1977	Protection Against Low-Trajectory Turbine Missles
1.116 1977	Quality Assurance Requirements for Installation, Inspection, and Testing of Mechanical Equipment and Systems
1.117 1978	Tornado Design Classification
1.118 1978	Periodic Testing of Electric Power and Protection Systems
1.120 1977	Fire Protection Guidelines for Nuclear Power Plants
1.121 1976	Bases for Plugging Degraded PWR Steam Generator Tubes
1.122 1978	Development of Floor Design Response Spectra for Seismic Design of Floor-Supported Equipment or Components
1.123 1977	Quality Assurance Requirements for Control of Procurement of Items and Services for Nuclear Power Plants

Nuclear Regulatory Guides for Power Reactors (Continued)

1.124 1978	Service Limits and Loading Combinations for Class 1 Linear-Type Component Supports
1.125 1978	Physical Models for Design and Operation of Hydraulic Structures and Systems for Nuclear Power Plants
1.126 1978	An Acceptable Model and Related Statistical Methods for the Analysis of Fuel Densification
1.127 1978	Inspection of Water-Cooled Structures Associated with Nuclear Power Plants
1.128 1978	Installation Design and Installation of Large Lead Storage Batteries for Nuclear Power Plants
1.129 1978	Maintenance, Testing, and Replacement of Large Lead Storage Batteries for Nuclear Power Plants
1.130 1978	Service Limits and Loading Combinations for Class 1 Plate-and-Shell-Type Component Supports
1.132 1979	Site Investigations for Foundations of Nuclear Power Plants
1.133 1981	Loose-Part Detection Program for the Primary Systems of Light-Water-Cooled Reactors
1.134 1979	Medical Evaluation of Nuclear Power Plant Personnel Requiring Operator License
1.135 1977	Normal Water Level and Discharge at Nuclear Power Plants
1.136 1981	Materials, Construction and Testing of Concrete Containments (Articles CC-1000, -2000, and -4000 through -6000 of the ACI "Code for Concrete Reactor Vessels and Containments")
1.137 1979	Fuel-Oil Systems for Standby Diesel Generators
1.138 1978	Laboratory Investigations of Soils for Engineering Analysis and Design of Nuclear Power Plants
1.139 1978	Guidance for Residual Heat Removal
1.140 1979	Design, Testing, and Maintenance Criteria for Normal Ventilation Exhaust System Air Filtration and Adsorption Units of Light-Water-Cooled Nuclear Power Plants
1.141 1978	Containment Isolation Provisions for Fluid Systems
1.142 1981	Safety-Related Concrete Structures for Nuclear Power Plants (Other Than Reactor Vessels and Containments)
1.143 1979	Design Guidance for Radioactive Waste Management Systems, Structures, and Components Installed in Light-Water-Cooled Nuclear Power Plants
1.144 1980	Auditing of Quality Assurance Programs for Nuclear Power Plants
1.145 1983	Atmospheric Dispersion Models for Potential Accident Consequence Assessments at Nuclear Power Plants
1.146 1980	Qualification of Quality Assurance Program Audit Personnel for Nuclear Power Plants
1.147 1983	Inservice Inspection Code Case Acceptability—ASME Section XI, Division 1

Nuclear Regulatory Guides for Power Reactors (Continued)

1.148 1981	Functional Specification for Active Valve Assemblies in Systems Important to Safety in Nuclear Power Plants
1.149 1981	Nuclear Power Plant Simulators in Operating Training
1.150 1983	Ultrasonic Testing of Reactor Vessel Welds During Preservice and Inservice Examinations
1.151 1983	Instrument Sensing Lines

Nuclear Regulatory Guides for Research and Test Reactors

GUIDE/YR	REGULATORY GUIDES, RESEARCH AND TEST REACTORS
2.1 1973	Shield Test Program for Evaluation of Installed Biological Shielding in Research and Training Reactors
2.2 1973	Development of Technical Specifications for Experiments in Research Reactors
2.3 1976	Quality Verification for Plate-Type Uranium–Aluminum Fuel Elements for Use in Research Reactors
2.4 1977	Review of Experiments for Research Reactors
2.5 1977	Quality Assurance Program Requirements for Research Reactors
2.6 1979	Emergency Planning for Research and Test Reactors

Nuclear Regulatory Guides for Fuels and Materials Facilities

GUIDE/YR	REGULATORY GUIDES, FUELS AND MATERIALS FACILITIES
3.1 1982	Use of Borosilicate-Glass Raschig Rings as a Neutron Absorber in Solutions of Fissile Material
3.2 1973	Efficiency Testing of Air-Cleaning Systems Containing Devices for Removal of Particles
3.3 1974	Quality Assurance Program Requirements for Fuel Reprocessing Plants and for Plutonium Processing and Fuel Fabrication Plants
3.4 1978	Nuclear Criticality Safety in Operations with Fissionable Materials Outside Reactors
3.5 1977	Standard Format and Content of License Applications for Uranium Mills
3.6 1973	Content of Technical Specifications for Fuel Reprocessing Plants

Nuclear Regulatory Guides for Fuels and Materials Facilities (Continued)

3.7 1973	Monitoring of Combustible Gases and Vapors in Plutonium Processing and Fuel Fabrication Plants
3.8 1982	Preparation of Environmental Reports for Uranium Mills
3.9 1973	Concrete Radiation Shields
3.10 1973	Liquid Waste Treatment System Design Guide for Plutonium Processing and Fuel Fabrication Plants
3.11 1973	Design, Construction, and Inspection of Embankment Retention Systems for Uranium Mills
3.11.1 1980	Operational Inspection and Surveillance of Embankment Retention Systems for Uranium Mill Tailings
3.12 1973	General Design Guide for Ventilation Systems of Plutonium Processing and Fuel Fabrication Plants
3.13 1973	Guide for Acceptable Waste Storage Methods at UF_6 Production Plants
3.14 1973	Seismic Design Classification for Plutonium Processing and Fuel Fabrication Plants
3.15 1983	Standard Format and Content of License Applications for Storage Only of Unirradiated Reactor Fuel and Associated Radioactive Material
3.16 1974	General Fire Protection Guide for Plutonium Processing and Fuel Fabrication Plants
3.17 1974	Earthquake Instrumentation for Fuel Reprocessing Plants
3.18 1974	Confinement Barriers and Systems for Fuel Reprocessing Plants
3.19 1974	Reporting of Operating Information for Fuel Reprocessing Plants
3.20 1974	Process Offgas Systems for Fuel Reprocessing Plants
3.21 1974	Quality Assurance Requirements for Protective Coatings Applied to Fuel Reprocessing and to Plutonium Processing and Fuel Fabrication Plants
3.22 1974	Periodic Testing of Fuel Reprocessing Plant Protection System Actuation Functions
3.25 1974	Standard Format and Content of Safety Analysis Reports for Uranium Enrichment Facilities
3.26 1975	Standard Format and Content of Safety Analysis Reports for Fuel Reprocessing Plants
3.27 1977	Nondestructive Examination of Welds in the Liners of Concrete Barriers in Fuel Reprocessing Plants
3.28 1975	Welder Qualification for Welding in Areas of Limited Accessibility in Fuel Reprocessing Plants and in Plutonium Processing and Fuel Fabrication Plants
3.29 1975	Preheat and Interpass Temperature Control for the Welding of Low-Alloy Steel for Use in Fuel Reprocessing Plants and in Plutonium Processing and Fuel Fabrication Plants
3.30 1975	Selection, Application, and Inspection of Protective Coatings (Paints) for Fuel Reprocessing Plants

Nuclear Regulatory Guides for Fuels and Materials Facilities (Continued)

3.31 1977	Emergency Water Supply Systems for Fuel Reprocessing Plants
3.32 1975	General Design Guide for Ventilation Systems for Fuel Reprocessing Systems
3.33 1977	Assumptions Used for Evaluating the Potential Radiological Consequences of Accidental Nuclear Criticality in a Fuel Reprocessing Plant
3.34 1979	Assumptions Used for Evaluating the Potential Radiological Consequences of Accidental Nuclear Criticality in a Uranium Fuel Fabrication Plant
3.35 1979	Assumptions Used for Evaluating the Potential Radiological Consequences of Accidental Nuclear Criticality in a Plutonium Processing and Fuel Fabrication Plant
3.37 1975	Guidance for Avoiding Intergranular Corrosion and Stress Corrosion in Austenitic Stainless Steel Components of Fuel Reprocessing Plants
3.38 1976	General Fire Protection Guide for Fuel Reprocessing Plants
3.39 1976	Standard Format and Content of License Applications for Plutonium Processing and Fuel Fabrication Plants
3.40 1977	Design Basis Floods for Fuel Reprocessing Plants and for Plutonium Processing and Fuel Fabrication Plants
3.41 1977	Validation of Calculational Methods for Nuclear Criticality Safety
3.42 1979	Emergency Planning for Fuel Cycle Facilities and Plants Licensed Under 10 CFR Parts 50 and 70
3.43 1979	Nuclear Criticality Safety in the Storage of Fissile Materials
3.44 1980	Standard Format and Content for the Safety Analysis Report for an Independent Spent Fuel Storage Installation (Water-Basin Type)
3.45 1980	Nuclear Criticality Safety for Pipe Intersections Containing Aqueous Solutions of Enriched Uranyl Nitrate
3.46 1982	Standard Format and Content of License Applications, Including Environmental Reports, for in Situ Uranium Solution Mining
3.47 1981	Nuclear Criticality Control and Safety of Homogeneous Plutonium–Uranium Fuel Mixtures Outside Reactors
3.48 1981	Standard Format and Content for the Safety Analysis Report for an Independent Spent Fuel Storage Installation (Dry Storage)
3.49 1982	Design of an Independent Spent Fuel Storage Installation (Water Basin Type)
3.50 1982	Guidance on Preparing a License Application to Store Spent Fuel in an Independent Spent Fuel Storage Installation
3.51 1982	Calculational Models for Estimating Radiation Doses to Man from Airborne Radioactive Materials Resulting from Uranium Milling Operations
3.52 1982	Standard Format and Content for the Health and Safety Sections of License Renewal Applications for Uranium Fuel Fabrication Plants
3.53 1982	Applicability of Existing Regulatory Guides to the Design and Operation of Independent Spent Fuel Storage Installation

Nuclear Regulatory Guides for Environmental and Siting Matters

GUIDE/YR	REGULATORY GUIDES, ENVIRONMENTAL AND SITING MATTERS
4.1 1975	Programs for Monitoring Radioactivity in the Environs of Nuclear Power Plants
4.2 1976	Preparation of Environmental Reports for Nuclear Power Stations
4.4 1974	Reporting Procedure for Mathematical Models Selected to Predict Heated Effluent Dispersion in Natural Water Bodies
4.5 1974	Measurements of Radionuclides in the Environment—Sampling and Analysis of Plutonium in Soil
4.6 1974	Measurements of Radionuclides in the Environment—Strontium-89 and Strontium-90 Analyses
4.7 1975	General Site Suitability Criteria for Nuclear Power Stations
4.8 1975	Environmental Technical Specifications for Nuclear Power Plants
4.9 1975	Preparation of Environmental Reports for Commercial Uranium Enrichment Facilities
4.11 1977	Terrestrial Environmental Studies for Nuclear Power Stations
4.13 1977	Performance, Testing, and Procedural Specifications for Thermoluminescence Dosimetry: Environmental Applications
4.14 1980	Radiological Effluent and Environmental Monitoring at Uranium Mills
4.15 1979	Quality Assurance for Radiological Monitoring Programs (Normal Operations)—Effluent Streams and the Environment
4.16 1978	Measuring, Evaluating, and Reporting Radioactivity in Releases of Radioactive Materials in Liquid and Airborne Effluents from Nuclear Fuel Processing and Fabrication Plants
4.17 1982	Standard Format and Content of Site Characterization Reports for High-Level-Waste Geologic Repositories
4.18 1983	Standard Format and Content of Environmental Reports for Near-Surface Disposal of Radioactive Waste

Nuclear Regulatory Guides for Materials and Plant Protection

GUIDE/YR	REGULATORY GUIDES, MATERIALS AND PLANT PROTECTION
5.1 1972	Serial Numbering of Fuel Assemblies for Light-Water-Cooled Power Reactors
5.3 1973	Statistical Terminology and Notation for Special Nuclear Materials Control and Accountability
5.4 1973	Standard Analytical Methods for the Measurement of Uranium Tetrafluoride (UF_4) and Uranium Hexafluoride (UF_6)
5.5	Standard Methods for Chemical, Mass Spectrometric, and

Nuclear Regulatory Guides for Materials and Plant Protection (Continued)

1973	Spectrochemical Analysis of Nuclear-Grade Uranium Dioxide Powders and Pellets
5.6 1973	Standard Methods for Chemical, Mass Spectrometric, and Spectrochemical Analysis of Nuclear-Grade Plutonium Dioxide Powders and Pellets and Nuclear-Grade Mixed Oxides, $(U,Pu)O_2$
5.7 1980	Entry/Exit Control for Protected Areas, Vital Areas, and Material Access Areas
5.8 1974	Design Considerations for Minimizing Residual Holdup of Special Nuclear Material in Drying and Fluidized Bed Operations
5.9 1974	Specifications for Ge(Li) Spectroscopy Systems for Material Protection Measurements—Part 1: Data Acquisition Systems
5.10 1973	Selection and Use of Pressure-Sensitive Seals on Containers for Onsite Storage of Special Nuclear Material
5.11 1973	Nondestructive Assay of Special Nuclear Material Contained in Scrap and Waste
5.12 1973	General Use of Locks in the Protection and Control of Facilities and Special Nuclear Materials
5.13 1973	Conduct of Nuclear Material Physical Inventories
5.14 1980	Use of Observation (Visual Surveillance) Techniques in Material Access Areas
5.15 1974	Security Seals for the Protection and Control of Special Nuclear Material
5.16 1975	Standard Methods for Chemical, Mass Spectrometric, Spectrochemical, Nuclear and Radiochemical Analysis of Nuclear-Grade Plutonium Nitrate Solutions and Plutonium Metal
5.17 1974	Truck Identification Markings
5.18 1974	Limit of Error Concepts and Principles of Calculation in Nuclear Materials Control
5.19 1974	Methods for the Accountability of Plutonium Nitrate Solutions
5.20 1974	Training, Equipping, and Qualifying of Guards and Watchmen
5.21 1974	Nondestructive Uranium-235 Enrichment Assay by Gamma-Ray Spectrometry
5.22 1974	Assessment of the Assumption of Normality (Employing Individual Observed Values)
5.23 1974	In Situ Assay of Plutonium Residual Holdup
5.24 1974	Analysis and Use of Process Data for the Protection of Special Nuclear Material
5.25 1974	Design Considerations for Minimizing Residual Holdup of Special Nuclear Material in Equipment for Wet Process Operations
5.26 1975	Selection of Material Balance Areas and Item Control Areas
5.27 1974	Special Nuclear Material Doorway Monitors

Nuclear Regulatory Guides for Materials and Plant Protection (Continued)

5.28 1974	Evaluation of Shipper–Receiver Differences in the Transfer of Special Nuclear Materials
5.29 1975	Nuclear Material Control Systems for Nuclear Power Plants
5.30 1974	Materials Protection Contingency Measures for Uranium and Plutonium Fuel Manufacturing Plants
5.31 1975	Specially Designed Vehicle with Armed Guards for Road Shipment of Special Nuclear Material
5.32 1975	Communication with Transport Vehicles
5.33 1974	Statistical Evaluation of Material Unaccounted for
5.34 1974	Nondestructive Assay for Plutonium in Scrap Material by Spontaneous Fission Detection
5.36 1974	Recommended Practice for Dealing with Outlying Observations
5.37 1983	In Situ Assay of Enriched Uranium Residual Holdup
5.38 1983	Nondestructive Assay of High-Enrichment Uranium Plates by Gamma-Ray Spectrometry
5.39 1974	General Methods for the Analysis of Uranyl Nitrate Solutions for Assay, Isotopic Distribution, and Impurity Determinations
5.40 1974	Methods for the Accountability of Plutonium Dioxide Powder
5.42 1975	Design Considerations for Minimizing Residual Holdup of Special Nuclear Material in Equipment for Dry Process Operations
5.43 1975	Plant Security Force Duties
5.44 1980	Perimeter Intrusion Alarm Systems
5.45 1974	Standard Format and Content for the Special Nuclear Material Control and Accounting Section of a Special Nuclear Material License Application (Including That for a Uranium Enrichment Facility)
5.47 1975	Control and Accountability of Plutonium in Waste Material
5.48 1975	Design Considerations—Systems for Measuring the Mass of Liquids
5.49 1975	Internal Transfers of Special Nuclear Materials
5.51 1975	Management Review of Nuclear Material Control and Accounting Systems
5.52 1980	Standard Format and Content of a Licensee Physical Protection Plan for Strategic Special Nuclear Material at Fixed Sites (Other Than Nuclear Power Plants)
5.53 1975	Qualification, Calibration, and Error Estimation Methods for Nondestructive Assay
5.54 1978	Standard Format and Content of Safeguards Contingency Plans for Nuclear Power Plants
5.55 1978	Standard Format and Content of Safeguards Contingency Plans for Fuel Cycle Facilities

Nuclear Regulatory Guides for Materials and Plant Protection (Continued)

5.56 1978	Standard Format and Content of Safeguards Contingency Plans for Transportation
5.57 1980	Shipping and Receiving Control of Strategic Special Nuclear Material
5.58 1980	Considerations for Establishing Traceability of SNM Accounting Measurements
5.59 1980	Standard Format and Content for a Licensee Physical Security Plan for the Protection of Special Nuclear Material of Moderate or Low Strategic Significance
5.60 1980	Standard Format and Content of a Licensee Physical Protection Plan for Strategic Special Nuclear Material in Transit
5.61 1980	Intent and Scope of the Physical Protection Upgrade Rule Requirements for Fixed Sites
5.62 1981	Reporting of Physical Security Events
5.63 1982	Physical Protection for Transient Shipments

Nuclear Regulatory Guides for Radioactive Products

GUIDE/YR	REGULATORY GUIDES, PRODUCTS (RADIOACTIVE)
6.1 1974	Leak Testing Radioactive Brachytherapy Sources
6.2 1974	Integrity and Test Specifications for Selected Brachytherapy Sources
6.3 1974	Design, Construction, and Use of Radioisotopic Power Generators for Certain Land and Sea Applications
6.4 1980	Classification of Containment Properties of Sealed Radioactive Sources
6.5 1974	General Safety Standard for Installations Using Nonmedical Sealed Gamma-Ray Sources
6.6 1974	Acceptance Sampling Procedures for Exempted and Generally Licensed Items Containing Byproduct Material
6.7 1976	Preparation of an Environmental Report To Support a Rule Making Petition Seeking an Exemption for Radionuclide-Containing Product
6.8 1979	Identification for Irretrievable Well-Logging Sources

Nuclear Regulatory Guides for Transportation

GUIDE/YR	REGULATORY GUIDES FOR TRANSPORTATION
7.1 1974	Administrative Guide for Packaging and Transporting Radioactive Material
7.2 1974	Packaging and Transportation of Radioactively Contaminated Biological Materials

Nuclear Regulatory Guides for Transportation (Continued)

7.3 1975	Procedures for Picking Up and Receiving Packages of Radioactive Material
7.4 1975	Leakage Tests on Packages for Shipment of Radioactive Materials
7.5 1977	Administrative Guide for Obtaining Exemptions From Certain NRC Requirements Over Radioactive Material Shipments
7.6 1978	Design Criteria for the Structural Analysis of Shipping Cask Containment Vessels
7.7 1977	Administrative Guide for Verifying Compliance With Packaging Requirements for Shipments of Radioactive Materials
7.8 1977	Load Combinations for the Structural Analysis of Shipping Casks
7.9 1980	Standard Format and Content of Part 71 Applications for Approval of Packaging of Type B, Large Quantity, and Fissile Radioactive Material
7.10 1983	Establishing Quality Assurance Programs for Packaging Used in the Transport of Radioactive Material

Nuclear Regulatory Guides for Occupational Health

GUIDE/YR	REGULATORY GUIDES, OCCUPATIONAL HEALTH
8.1 1973	Radiation Symbol
8.2 1973	Administrative Practices in Radiation Monitoring
8.3 1973	Film Badge Performance Criteria
8.4 1973	Direct-Reading and Indirect-Reading Pocket Dosimeters
8.5 1981	Criticality and Other Interior Evacuation Signals
8.6 1973	Standard Test Procedure for Geiger-Muller Counters
8.7 1973	Occupational Radiation Exposure Records Systems
8.8 1978	Information Relevant to Ensuring That Occupational Radiation Exposures at Nuclear Power Stations Will be As Low As Reasonably Achievable (ALARA)
8.9 1973	Acceptable Concepts, Models, Equations, and Assumptions for a Bioassay Program
8.10 1977	Operating Philosophy for Maintaining Occupational Radiation Exposures As Low As Is Reasonably Achievable (Nuclear Power Reactors)
8.11 1974	Applications of Bioassay for Uranium
8.12 1981	Criticality Accident Alarm Systems

Nuclear Regulatory Guides for Occupational Health (Continued)

8.13 1975	Instruction Concerning Prenatal Radiation Exposure
8.14 1977	Personnel Neutron Dosimeters
8.15 1976	Acceptable Programs for Respiratory Protection
8.18 1982	Information Relevant to Ensuring That Occupational Radiation Exposures at Medical Institutions Will Be As Low As Reasonably Achievable
8.19 1979	Occupational Radiation Dose Assessment in Light-Water Reactor Power Plants—Design Stage Man-Rem Estimates
8.20 1979	Applications for Bioassay for I-125 and I-131
8.21 1979	Health Physics Surveys for Byproduct Material at NRC-Licensed Processing and Manufacturing Plants
8.22 1978	Bioassay at Uranium Mills
8.23 1981	Radiation Safety Surveys at Medical Institutions
8.24 1981	Health Physics Surveys During Enriched Uranium-235 Processing and Fuel Fabrication
8.25 1980	Calibration and Error Limits of Air Sampling Instruments for Total Volume Air Sampled
8.26 1980	Applications of Bioassay for Fission and Activation Products
8.27	Radiation Protection Training for Personnel at Light-Water-Cooled Nuclear Power Plants
8.28 1981	Audible-Alarm Dosimeters
8.29 1981	Instruction Concerning Risks from Occupational Radiation Exposure
8.30 1983	Health Physics Surveys in Uranium Mills
8.31 1983	Information Relevant to Ensuring that Occupational Radiation Exposures at Uranium Mills Will Be As Low As Is Reasonably Achievable

Nuclear Regulatory Guides for Antitrust and Financial Review

GUIDE/YR	REGULATORY GUIDES, ANTITRUST AND FINANCIAL REVIEW
9.1 1973	Regulatory Staff Position Statement on Antitrust Matters
9.2 1976	Information Needed by the NRC Staff in Connection with Its Antitrust Review of Construction Permit Applications for Nuclear Power Plants

Nuclear Regulatory Guides for Antitrust and Financial Review (Continued)

9.3 1974	Information Needed by the AEC Regulatory Staff in Connection with Its Antitrust Review of Operating License Applications for Nuclear Power Plants
9.4 1979	Suggested Format for Cash Flow Statements Submitted as Guarantees of Payment of Retrospective Premiums

Nuclear Regulatory Guides, General

GUIDE/YR	REGULATORY GUIDES, PREPARATION OF APPLICATIONS
10.1 1981	Compilation of Reporting Requirements for Persons Subject to NRC Regulations
10.2 1976	Guidance to the Academic Institutions Applying for Specific Byproduct Material Licenses of Limited Scope
10.3 1977	Guide for the Preparation of Applications for Special Nuclear Material Licenses of Less Than Critical Mass Quantities
10.4 1977	Guide for the Preparation of Applications for Licenses to Process Source Material
10.5 1981	Applications for Type A Licenses of Broad Scope
10.6 1981	Guide for the Preparation of Applications for Use of Sealed Sources and Devices for Performing Industrial Radiography
10.7 1979	Guide for the Preparation of Applications for Licenses for Laboratory Use of Small Quantities of Byproduct Material
10.8 1980	Guide for the Preparation of Applications for Medical Programs
10.9 1980	Guide for the Preparation of Applications for Licenses for the Use of Gamma Irradiators

9.2 SERIALS ISSUED BY GOVERNMENTAL BODIES

Serials of Former U.S. Atomic Energy Commission

The topical content of reports listed below, issued by the former U.S. Atomic Energy Commission, is now covered by the reports and publications of the Nuclear Regulatory Commission or the Department of Energy.

Publication	Volume & Period
Reactor and Fuel Processing Technology (cont. as *Reactor Materials*)	11–12, 1968–1970
Reactor Core Materials (cont. as *Reactor Materials*)	1–4, 1958–1961

Publication	Volume & Period
Reactor Fuel Processing (cont. as *Reactor and Fuel-Processing Technology*)	1–9, 1958–1966
Reactor Materials (formerly *Reactor Core Materials*)	5–13, 1962–1970
Reactor Technology (formerly *Reactor Materials*, and *Reactor and Fuel Processing Technology*)	13–15, 1970–1972
U.S. AEC Serious Accidents	1973–1975
Isotopes and Radiation Technology	1–9, 1963–1972
Power Reactor Technology (cont. as *Reactor and Fuel Processing Technology*)	1–9, 1957–1966
AEC-NASA Tech Briefs	1965–1972

Nuclear Regulatory Commission

The Nuclear Regulatory Commission publishes a large body of documents under the NUREG category, which report contract work carried out on behalf of the Commission.

The titles of these documents are as follows:

- NUREG INDEX, see Regulatory and Technical Reports
- NUREG-0020, see Licensed Operating Reactors, Status Summary Report
- NUREG-0030, see Nuclear Power Plant Construction Status Report
- NUREG-0040, see Licensee Contractor and Vendor Inspection Status Report
- NUREG-0090, see Report to Congress on Abnormal Occurrences
- NUREG-0304, see Regulatory and Technical Reports
- NUREG-0390, see Topical Report Review Status
- NUREG-0485, see Systematic Evaluation Program Status Summary Report
- NUREG-0540, see Title List of Documents Made Publicly Available
- NUREG-0606, see Unresolved Safety Issues Summary
- NUREG-0800, see Standard Review Plan for the Review of Safety
- NUREG/CR-2000,see Licensee Event Reports

The most frequently consulted reports in the above series are: the monthly status summary report for licensed operating reactors, NUREG-0020, otherwise known as the "Graybook," and a quarterly report for nuclear power plants currently under construction, NUREG-0030, known as the "Yellowbook". A newsletter published by McGraw-Hill under the title "Inside N.R.C." informs those interested of the Commission's activities.

Nuclear Safety, published by the Nuclear Regulatory Commission, reports developments in the area of reactor safety.

Publication	Volume & Period
Standard Review Plan for the Review of Safety Analysis Reports for Nuclear Power Plants	1975–
Standards Development Status Summary Report	1981–
Systematic Evaluation Program Status Summary Report	3– , 1981–
Title List of Documents Made Publicly Available (formerly: Power Reactor Docket Information)	1– , 1979–
Topical Report Review Status	1981–
Regulatory and Technical Reports	3– , 1975–
Regulatory Guides	1970–
Licensed Operating Reactors, Status Summary Report	1974–1983
Licensee Contractor and Vendor Inspection Status Report	1–6, 1976–1982
Licensee Event Reports	1980–
Nuclear Power Plant Construction Status Report	1975–1983
Issuances	1– , 1975–
Press Releases	1975–
IE (Inspection and Enforcement) Bulletins	1971–
IE Circulars	1976–
IE Information Notices	1979–
Operating Experience Bulletin Information Report	1974–1979
Power Reactors—Current Events (continued as: Power Reactor Events)	1974–1978
Power Reactor Events (formerly: Power Reactors—Current Events)	1– , 1979–
Report to Congress on Abnormal Occurrences	1–4, 1976–1980

Department of Energy

The Department of Energy publishes two series of "Current Awareness Bulletins": *Nuclear Fuel Cycle,* and *Radioactive Waste Management,* the first appearing in 24 issues per year. It also publishes a newsletter, "Inside Energy" (formerly "Inside D.O.E.").

Publication	Volume & Period
Monthly Energy Review	
Department of Energy Information and Data	1– , 1977–
Energy Information Directory	
Nuclear Safety	1–, 1959–
Power Reactor Docket Information	1974–1978
Radioactive Waste Technology Program Newsletter	current year

Other Government Departments

U.S. Environmental Protection Agency:	Volume & Period
Radiological Health Data and Reports (cont. as *Radiation Data and Reports*)	6–12, 1965–1971
Radiation Data and Reports, (formerly *Radiological Health Data and Reports*)	13–15, 1972–1974
Government Publishing Office:	
Code of Federal Regulations	current year
List of CFR Sections Affected	current
Federal Register	25– , 1960–
U.S. Congress:	
Congressional Record Index	1978–
U.S. Department of Labor:	
Consumer Prices: Energy	current year
U.S. Department of Interior:	
Earthquake Information Bulletin	4– , 1972–

Intergovernmental Bodies

The International Atomic Energy Agency in Vienna, Austria publishes an annual serial, "Operating Experience with Nuclear Power Stations in Member States," a catalog of reactor design and operating data, IAEA-127, as well as a "Performance Analysis Report" under the same general title. The Agency is also a major publisher of proceedings of topical conferences, reviews and monographs on subjects connected with nuclear energy, the fuel cycle, environmental safety and others. The publications of IAEA have by now grown into a respectable library, covering the entire field of nuclear engineering and technology, as well as spent fuel handling and reprocessing.

International Atomic Energy Agency:	Volume & Period
Atomic Energy Review	1–18, 1963–1980
INIS Atomindex	1—7, 1970–1976
Bulletin	1–20, 1959–1978
Meetings on Atomic Energy	7–13, 1975–1981
Nuclear Fusion	12–14, 1972–1974
Waste Energy Research Abstracts	1968–

Other International Bodies

In recent times the Organization for Economic Cooperation and Development has entered the field and started publishing monographs on the nuclear fuel cycle, radioactive waste management, law with relation to nuclear installations, and radiation protection.

Publications	Volume & Period
European Community:	
Euratom Quarterly Digest	1–3, 1960–1963
Euratom Information (cont. as Euro Abstracts)	1–7, 1963–1969
Euro Abstracts	8–11, 1970–1973
Euratom Bulletin	2–6, 1963–1967
European Atomic Energy Community:	
Transatom Bulletin	1–5, 1960–1965

Abstracting Services

The principal abstracting service in the area of nuclear engineering between 1948 and 1976 was *"Nuclear Science Abstracts,"* published by the then-*U.S. Atomic Energy Commission.* Altogether 33 volumes of abstracts, covering the entire field of nuclear engineering and science, were produced.

After 1976 the publication of abstracts in nuclear science was taken over by the *International Atomic Energy Agency in Vienna, Austria.* The Agency publishes the *"Atomindex,"* modeled very closely on the former *"Nuclear Science Abstracts."* At this writing the publication is in its 15-th volume (1984).

"The Nuclear Index," published by the *National Energy Researchers,* lists the current titles in the nuclear literature.

"The Engineerng Index" covers reports and articles dealing with the engineering aspects of nuclear science.

	Volume & Period
Energy Research Abstracts	1– , 1976–
U.S. Department of Energy	
Engineering Index Annual	72–81, 1973–1982
Engineering Information	
Engineering Index Monthly	19– , 1982–
Engineering Information	
Nuclear Index	1– , 1980–
National Energy Researchers	
Nuclear Science Abstracts	1–33, 1947–1976
Energy Research and Development Administration	
Computer Program Abstracts	1–12, 1969–1981
National Aeronautics and Space Administration	

BIBLIOGRAPHY, BOOKS AND MONOGRAPHS

Reactor Materials

W. Kermit Anderson and J. S. Theilacker, eds., "Neutron Absorber Materials for Reactor Control," U. S. Atomic Energy Commission, Naval Reactors, Division of Reactor Development, 1962.

W. E. Berry, *Corrosion in Nuclear Applications,* John Wiley & Sons, New York, 1971.

Robert O. Bolt and James G. Carroll, eds., *Radiation Effects on Organic Materials,* Academic Press, New York, 1963.

Lewis T. Chadderton, *Radiation Damage in Crystals,* John Wiley & Sons, New York, 1965.

Lewis T. Chadderton and Ian M. Torrens, *Fission Damage in Crystals,* Methuen and Co., London, 1969.

J. M. Cleveland, *The Chemistry of Plutonium* American Nuclear Society, La Grange Park, IL, 1979.

Paul Cohen, *Water Coolant Technology of Power Reactors,* Gordon & Breach, Science Publishers, New York, 1969.

George A. Freund, *Materials for Control Rod Drive Mechanisms,* Rowman and Littlefield, New York, 1963.

J. H. Gittus, *Uranium,* Vol. 8 in *Metallurgy of the Rarer Metals,* Butterworth's, Bethlehem, MD, 1963.

Henry H. Hausner, ed., *Beryllium, Its Metallurgy and Properties,* University of California Press, Berkeley, CA, 1965.

George J. Janz, *Molten Salts Handbook,* Academic Press, New York, 1967.

B. T. Kelly, *Radiation Damage to Solids,* Pergamon Press, Elmsford, NY, 1966.

J. F. Kircher and R. E. Bowman, eds., *Effects of Radiation on Materials and Components,* Reinhold Publishers, New York, 1964.

R. F. Makens, ed., "Organic Coolant Summary Report," U. S. Atomic Energy Commission, Washington, DC, 1964.

S. S. Manson, *Thermal Stress and Low-Cycle Fatigue,* McGraw-Hill, New York, 1967.

R. D. Nightingale, ed., *Nuclear Graphite,* Academic Press, New York, 1962.

S. S. Penner, ed., *Lithium—Needs and Resources,* Pergamon Press, Elmsford, NY, 1978.

William E. Ray, *Fabrication of Control Rods for Nuclear Reactors,* Rowman and Littlefield, New York, 1963.

J. T. A. Roberts, *Structural Materials in Nuclear Power Systems,* Plenum Press, New York, 1981.

M. T. Simand and L. R. Zumwalt, eds., *Materials and Fuels for High-Temperature Nuclear Energy Applications,* MIT Press, Cambridge, MA, 1964.

H. W. Simmons, *Radiation Effects in Graphite,* Pergamon Press, Oxford, 1965.

Charles O. Smith, *Nuclear Reactor Materials,* Addison-Wesley Publishing Co., Reading, MA, 1967.

Donald E. Thomas and Earl T. Hayes, eds., "The Metallurgy of Hafnium," U.S. Government Printing Office, Washington, DC, 1960.

E. J. Wheelwright, ed., *Promethium Technology,* American Nuclear Society, La Grange Park, IL, 1973.

J. F. Ziegler, *The Stopping Ranges of Ions in Matter,* 4 vols, Pergamon Press, Elmsford, NY, 1980.

Reactor Thermo-Hydraulics

Richard C. Bailie, *Energy Conversion Engineering,* Addison-Wesley Publishing Co., Reading, MA, 1978.

R. D. Blevins, *Flow Induced Vibrations,* Van Nostrand Reinhold Co., New York, 1977.

O. E. Dwyer, *Boiling Liquid–Metal Heat Transfer,* American Nuclear Society, La Grange Park, IL, 1976.

M. M. El-Wakil, *Nuclear Power Engineering,* McGraw-Hill, New York, 1962.

M. M. El-Wakil, *Nuclear Energy Conversion,* American Nuclear Society, La Grange Park, IL, 1982.

M. M. El-Wakil, *Nuclear Heat Transport,* American Nuclear Society, La Grange Park, IL, 1981.

Henri Fenech, ed., *Heat Transfer and Fluid Flow in Nuclear Systems,* Pergamon Press, Elmsford, NY, 1981.

D. D. Kalafati, *Thermodynamic Cycles of Nuclear Power Stations,* Daniel Davey & Co., New York, 1965.

R. T. Lahey, Jr. and F. T. Moody, *The Thermal-Hydraulics of a Boiling Water Nuclear Reactor,* American Nuclear Society, La Grange Park, IL, 1977.

Y. S. Tang, R. D. Coffield, and R. A. Markley, *Thermal Analysis of Liquid Metal Fast Breeder Reactors,* American Nuclear Society, La Grange Park, IL, 1978.

L. S. Tong, *Boiling Heat Transfer and Two-Phase Flow,* John Wiley & Sons, New York, 1965.

L. S. Tong and Joel Weisman, *Thermal Analysis of Pressurized Water Reactors,* American Nuclear Society, 1970.

G. B. Wallis, *One Dimensional Two-Phase Flow,* McGraw-Hill, New York, 1969.

R. H. S. Winterton, *Thermal Design of Nuclear Reactors,* Pergamon Press, Elmsford, NY, 1981.

Reactor Control and Safety

Milton Ash, *Nuclear Reactor Kinetics,* McGraw-Hill, New York, 1965.

Milton Ash, *Optimal Shutdown Control of Nuclear Reactors,* Academic Press, New York, 1966.

F. R. Farmer, ed., *Nuclear Reactor Safety,* Academic Press, New York, 1977.

Wesley K. Foell, *Small-Sample Reactivity Measurements in Nuclear Reactors,* American Nuclear Society, 1972.

John Graham, *Fast Reactor Safety,* Academic Press, New York, 1971.

E. Arnold Hakkila, ed., "Nuclear Safeguards Analysis," American Chemical Society, New York, 1978.

Robert J. Hansen, ed., *Seismic Design for Nuclear Power Plants,* MIT Press, Cambridge, MA, 1970.

Joseph M. Harrier, *Nuclear Reactor Control Engineering,* D. Van Nostrand, Princeton, NJ, 1963.

D. L. Hetrick, *Dynamics of Nuclear Reactors,* University of Chicago Press, Chicago, IL, 1971.

Harry H. Hummel and David Okrent, *Reactivity Coefficients in Large Fast Power Reactors,* Gordon & Breach, Science Publishers, New York, 1969.

Owen C. Jones, *Nuclear Reactor Safety Heat Transfer,* Hemisphere Publishing Corporation, Washington, DC, 1981.

G. Robert Keepin, *Physics of Nuclear Kinetics,* Addison-Wesley Publishing Co., Reading, MA, 1965.

Thomas W. Kerlin, *Frequency Response Testing in Nuclear Reactors,* Academic Press, New York, 1974.

Jeffrey Lewins, *Nuclear Reactor Kinetics and Control,* Pergamon Press, Elmsford, NY, 1978.

E. E. Lewis, *Nuclear Power Reactor Safety,* John Wiley & Sons, New York, 1977.

W. W. Lawrance, *Of Acceptable Risks: Science and the Determination of Safety,* William Kaufmann, Los Altos, CA, 1976.

Ronald R. Mohler and C. N. Shen, *Optimal Control of Nuclear Reactors,* Academic Press, New York, 1971.

Nicola Pacilio, "Reactor Noise Analysis in the Time Domain," U.S. Atomic Energy Commission, Washington, DC, 1969.

J. H. Rust and L. E. Weaver, *Nuclear Power Safety,* Pergamon Press, Elmsford, NY, 1977.

M. A. Schultz, *Control of Nuclear Reactors and Power Plants,* 2nd edit., McGraw-Hill, New York, 1961.

Weston M. Stacey, *Space-Time Nuclear Reactor Kinetics,* Academic Press, New York, 1969.

Joseph A. Thie, *Power Reactor Noise,* American Nuclear Society, La Grange Park, IL, 1981.

T. J. Thompson, and J. C. Beckerley, eds., *The Technology of Reactor Safety,* 2 vols., MIT Press, Cambridge, MA, 1964 and 1973.

Robert E. Uhrig, *Random Noise Techniques in Nuclear Reactor Systems,* Ronald Press Company, New York, 1970.

Lynn E. Weaver, *System Analysis of Nuclear Reactor Dynamics,* Rowman and Littlefield, New York, 1963.

M. M. R. Williams, *Random Processes in Nuclear Reactors,* Pergamon Press, Elmsford, NY, 1974.

M. M. R. Williams, *Nuclear Safety,* Pergamon Press, Elmsford, NY, 1979.

Fission Reactor Theory

George I. Bell and Samuel Glasstone, *Nuclear Reactor Theory,* Van Nostrand Reinhold Co., New York, 1970.

I. R. Cameron, *Nuclear Fission Reactors,* Plenum Publishing Co., New York, 1982.

Kenneth M. Case and P. F. Zweifel, *Linear Transport Theory,* Addison-Wesley Publishing Co., Reading, MA, 1967.

Thomas J. Connolly, *Foundations of Nuclear Engineering,* John Wiley & Sons, New York, 1978.

Lawrence Dresner, *Resonance Absorption in Nuclear Reactors,* Pergamon Press, Elmsford, NY, 1960.

James J. Duderstadt and Louis J. Hamilton, *Nuclear Reactor Analysis,* John Wiley & Sons, New York, 1976.

James Duderstadt and W. R. Martin, *Transport Theory,* John Wiley & Sons, New York, 1979.

J. H. Ferziger and P. F. Zweifel, *The Theory of Neutron Slowing Down in Nuclear Reactors,* Pergamon Press, Elmsford, NY, 1966.

Anthony Foderaro, *The Elements of Neutron Interaction Theory,* MIT Press, Cambridge, MA, 1971.

A. D. Galanin, *Thermal Reactor Theory,* 2nd edit., Pergamon Press, Elmsford, NY, 1960.

Samuel Glasstone and Milton C. Edlund, *The Elements of Nuclear Reactor Theory,* D. Van Nostrand, Princeton, NJ, 1952.

H. S. Isbin, *Introductory Nuclear Reactor Theory,* Reinhold Publishing Co., New York, 1963.

D. Jakeman, *Physics of Nuclear Reactors,* American Elsevier Publishing Co., New York, 1966.

John R. Lamarsh, *Introduction to Nuclear Reactor Theory,* Addison-Wesley Publishing Co., Reading, MA, 1966.

Jeffrey Lewins, *Importance: The Adjoint Function,* Pergamon Press, Elmsford, NY, 1965.

Jeffrey Lewins and Martin Becker, eds., *Sensitivity and Uncertainty Analysis of Reactor Performance Parameters,* Vol. 14, *Advances in Nuclear Science and Technology,* Plenum Publishing Co., New York, 1982.

S. E. Liverhant, *Elementary Introduction to Nuclear Reactor Physics,* John Wiley & Sons, New York, 1960.

J. E. Lynn, *Theory of Neutron Resonance Reactions,* Oxford University Press, New York, 1968.

Luigi Massimo, *Physics of High Temperature Reactors,* Pergamon Press, Oxford, 1975.

J. L. Meem, *Two-Group Reactor Theory,* Gordon & Breach, Science Publishers, New York, 1964.

Robert V. Meghreblian and David K. Holmes, *Reactor Analysis,* McGraw-Hill, New York, 1960.

Raymond L. Murray, *Nuclear Reactor Physics,* Prentice-Hall, Englewood Cliffs, NJ, 1957.

Richard K. Osborn and Sidney Yip, *The Foundations of Neutron Transport Theory,* Gordon & Breach, Science Publishers, New York, 1966.

A. Radkowsky, ed., "Naval Reactors Physics Handbook, Vol. I: Selected Basic Techniques," U.S. Atomic Energy Commission, Washington, DC, 1964.

Roger A. Rydin, *Nuclear Reactor Theory and Design,* University Publications, Blacksburg, VA, 1977.

J. H. Tait, *Neutron Transport Theory,* American Elsevier Publishing Company, New York, 1965.

John R. Taylor, *Scattering Theory,* John Wiley & Sons, New York, 1972.

G. Milton Wing, *An Introduction to Transport Theory,* John Wiley & Sons, New York, 1962.

L. C. Woods, *Introduction to Neutron Distribution Theory,* Methuen and Co., London, 1964.

P. F. Zweifel, *Reactor Physics,* McGraw-Hill, New York, 1973.

Reactor Physics Experimentation

K. H. Beckurts and K. Wirtz, *Neutron Physics,* Springer-Verlag, New York, 1964.

G. Dearnaley and D. C. Northrop, *Semiconductor Counters for Nuclear Radiations,* 2nd edit., John Wiley & Sons, New York, 1966.

P. A. Egelstaff, ed., *Thermal Neutron Scattering,* Academic Press, New York, 1965.

P. A. Egelstaff and M. J. Poole, *Experimental Neutron Thermalization,* Pergamon Press, Elmsford, NY, 1969.

R. D. Evans, *The Atomic Nucleus,* McGraw-Hill, New York, 1955.

J. A. Harvey, ed., *Experimental Neutron Resonance Spectroscopy,* Academic Press, New York, 1970.

J. B. Marion and D. M. Van Patter, *Nuclear Research with Low Energy Accelerators,* Academic Press, New York, 1967.

J. B. Marion and J. L. Fowler, *Fast Neutron Physics. Part I: Techniques,* Interscience Publishers, New York, 1960.

A. Edward Profio, *Experimental Reactor Physics,* John Wiley & Sons, New York, 1975.

Loren C. Schmid, *Critical Assemblies and Reactor Research,* John Wiley & Sons, New York, 1971.

Joseph A. Thie, *Heavy Water Exponential Experiments Using ThO_2 and UO_2,* Pergamon Press, Elmsford, NY, 1961.

Fission Reactor Engineering

D. J. Bennett, *The Elements of Nuclear Power,* Longman Group Ltd., London, 1972.

G. G. Eichholz, *Environmental Aspects of Nuclear Power,* Ann Arbor Science Publishers, Ann Arbor, MI, 1976.

Harold Etherington, ed., *Nuclear Engineering Handbook,* McGraw-Hill, New York, 1958.

A. R. Foster and R. L. Wright, *Basic Nuclear Engineering,* 3rd edit., Allyn and Bacon, Boston, MA, 1977.

B. R. T. Frost, *Nuclear Fuel Elements,* Pergamon Press, Elmsford, NY, 1982.

Samuel Glasstone and Walter H. Jordan, *Nuclear Power and Its Environmental Effects,* American Nuclear Society, La Grange Park, IL, 1980.

Samuel Glasstone and Alexander Sesonske, *Nuclear Reactor Engineering,* 3rd edit., Van Nostrand Reinhold Co., New York, 1981.

John F. Harvey, *Pressure Vessel Design: Nuclear and Chemical Applications,* D. Van Nostrand, Princeton, NJ, 1963.

C. D. Gregg King, *Nuclear Power Systems,* Macmillan, New York, 1964.

Ronald Allen Knief, *Nuclear Energy Technology,* McGraw-Hill, New York, 1981.

A. W. Kramer, *Boiling Water Reactors,* Addison-Wesley Publishing Co., Reading, MA, 1958.

L. Wang Lau, *Elements of Nuclear Reactor Engineering,* Gordon & Breach, Science Publishers, New York, 1974.

Kenneth C. Lish, *Nuclear Power Plant Systems and Equipment,* Industrial Press, New York, 1972.

Robert L. Loftness, *Nuclear Power Plants: Design, Operating Experience and Economics,* D. Van Nostrand, Princeton, NJ, 1964.

P. H. Margen, *Nuclear Reactor Optimization,* Simmons-Broadman, New York, 1960.

Glenn Murphy, *Elements of Nuclear Engineering,* John Wiley & Sons, New York, 1961.

Raymond L. Murray, *Introduction to Nuclear Engineering,* Prentice-Hall, Englewood Cliffs, NJ, 1961.

Raymond L. Murray, *Nuclear Energy,* 2nd edit., Pergamon Press, Elmsford, NY, 1980.

K. S. Ram, *Basic Nuclear Engineering,* Halsted Press, New York, 1977.

Reactor Handbook, Vol. I: *Materials,* C. R. Tipton, ed.; Vol. II: *Fuel Reprocessing,* M. S. Stoller and R. B. Richards, eds.; Vol. III-A: *Reactor Physics,* Harry Soodak, ed.; Vol. III-B: *Shielding,* Everitt Blizard and L. S. Abbott, eds.; Vol. IV: *Engineering,* Stuart McLain and John H. Martens, eds.; 2nd edit., Wiley-Interscience, New York, 1960–1964.

Henry Semat and John R. Albright, *Introduction to Atomic and Nuclear Physics,* 5th edit., Holt, Rinehart and Winston, New York, 1972.

Alexander Sesonske, "Nuclear Power Plant Design Analysis," TID-26241, U.S. Atomic Energy Commission, 1974.

J. Shaw, *Reactor Operation,* Pergamon Press, Elmsford, NY, 1969.

The Shippingport Pressurized Water Reactor, Addison-Wesley Publishing Co., Reading, MA, 1958.

Edward Teller, Wilson K. Talley, Gary Higgins, and Gerald W. Johnson, *The Constructive Uses of Nuclear Explosives,* McGraw-Hill, New York, 1968.

Robert Vandenbosch and John R. Huizenga, *Nuclear Fission,* Academic Press, New York, 1973.

Roy Weinstein, Alvin Boltax and Giovanni Lanza, *Nuclear Engineering Fundamentals,* McGraw-Hill, New York, 1964.

Joel Weisman, ed., *Elements of Nuclear Reactor Design,* American Elsevier Publishing Co., New York, 1977.

G. J. Wills, *Nuclear Power Plant Technology,* John Wiley & Sons, New York, 1967.

Fast Breeder Reactors

A. M. Judd, *Fast Breeder Reactors: An Engineering Introduction,* Pergamon Press, New York, 1981.

Alan E. Waltar and Albert B. Reynolds, *Fast Breeder Reactors,* Pergamon Press, New York, 1981.

Karl Wirtz, *Lectures on Fast Reactors,* American Nuclear Society, La Grange Park, IL, 1978.

John G. Yevick, *Fast Reactor Technology: Plant Design,* The MIT Press, Cambridge, MA, 1966.

S. Yiftah, D. Okrent, and P. A. Moldauer, *Fast Reactor Cross Sections,* Pergamon Press, Elmsford, NY, 1960.

Computational Methods

W. E. Alley and R. M. Lesler, *Neutron Activation Cross Sections, Measured and Semiempirical,* Nuclear Data Tables 11, Academic Press, New York, 1973.

W. W. Black, *An Introduction to On-Line Computers,* Gordon & Breach, Science Publishers, New York, 1971.

E. D. Cashwell and C. J. Everett, *A Practical Manual on the Monte Carlo Method for Random Walk Problems,* Pergamon Press, New York, 1959.

Melville Clark, Jr. and Kent F. Hansen, *Numerical Methods of Reactor Analysis,* Academic Press, New York, 1964.

Ruth N. Cuadra, David M. Abels, and Judith Wagner, eds., "Directory of Online Databases," Cuadra Associates, Inc., Santa Monica, CA, Vol. 3, No. 3, Spring 1982.

B. Davidson and J. B. Sykes, *Neutron Transport Theory,* Clarendon Press, Oxford, 1957.

Theodore E. Harris, *The Theory of Branching Processes,* Prentice-Hall, Englewood Cliffs, NJ, 1963.

John E. Hopcroft and Jeffrey D. Ullman, *Formal Languages and Their Relation to Automata,* Addison-Wesley Publishing Co., Reading, MA, 1969.

Donald E. Knuth, *The Art of Computer Programming,* 7 vols., Addison-Wesley Publishing Co., Reading, MA, 1969.

F. Wilfrid Lancaster, *Information Retrieval Systems Characteristics, Testing and Evaluation,* John Wiley & Sons, New York, 1968.

K. C. Peng, *The Design and Analysis of Scientific Experiments,* Addison-Wesley Publishing Co., Reading, MA, 1967.

Burt W. Rust and Walter R. Burrus, *Mathematical Programming and the Solution of Linear Equations,* American Elsevier Publishing Company, New York, 1972.

Ward C. Sangren, *Digital Computers and Nuclear Reactor Calculations,* Wiley, New York, 1960.

Akinao Shimizu and Katsutada Aoki, *Application of Invariant Imbedding to Reactor Physics,* Academic Press, New York, 1972.

Jerome Spanier and Ely M. Gelbard, *Monte Carlo Principles and Neutron Transport Problems,* Addison-Wesley Publishing Co., New York, 1969.

W. M. Stacey, *Modal Approximations: Theory and an Application to Reactor Physics,* The MIT Press, Cambridge, MA, 1967.

Weston M. Stacey, *Variational Methods in Nuclear Reactor Physics,* Academic Press, New York, 1974.

J. H. Tait, *Neutron Transport Theory,* American Elsevier Publishing Co., New York, 1965.

Eugene L. Wachspress, *Iterative Solution of Elliptic Systems and Application to the Neutron Diffusion Equations of Reactor Physics,* Prentice-Hall, Englewood Cliffs, NJ, 1966.

G. Milton Wing, *An Introduction to Transport Theory,* John Wiley & Sons, New York, 1962.

Finite Element Methods

K. J. Bathe and E. L. Wilson, *Numerical Methods in Finite Element Analysis,* Prentice-Hall, Englewood Cliffs, NJ, 1976.

T. Belytschko, P. V. Marcal, and J. R. Osias, eds., "Finite Element Analysis of Transient Nonlinear Structural Behavior," AMD-Vol. 14, American Society of Mechanical Engineers, New York, 1975.

Robert D. Cook, *Concepts and Applications of Finite Element Analysis,* John Wiley & Sons, New York, 1981.

C. S. Desai and John F. Abel, *Introduction to the Finite Element Method: A Numerical Method for Engineering Analysis,* Van Nostrand Reinhold, New York, 1972.

Richard H. Gallagher, *Finite Element Analysis: Fundamentals,* Civil Engineering and Engineering Mechanics Series, Prentice-Hall, Englewood Cliffs, NJ, 1975.

T. J. R. Hughes, ed., "Finite Element Methods for Convection Dominated Flows," AMD-Vol. 34, American Society of Mechanical Engineers, New York, 1979.

T. J. R. Hughes, ed., "Nonlinear Finite Element Analysis of Plates and Shells," AMD-Vol. 48, American Society of Mechanical Engineers, New York, 1981.

T. J. R. Hughes and D. Gartling, eds., "New Concepts in Finite Element Analysis," AMD-Vol. 44, American Society of Mechanical Engineers, New York, 1981.

M. M. Kamal and J. A. Wolf, "Finite Element Applications in Acoustics," American Association of Mechanical Engineers, New York, 1981.

M. S. Shepherd and R. H. Gallagher, "Finite Element Grid Optimization," PVP-Vol. 38, American Society of Mechanical Engineers, New York, 1979.

Eugene L. Wachspress, *A Rational Finite Element Basis,* Academic Press, New York, 1975.

O. C. Zienkiewicz, *The Finite Element Method in Engineering Science,* McGraw-Hill, London, 1971.

Fusion Reactors

Lev A. Artsimovich, *Elementary Plasma Physics,* Blaisdell Publishing Co., New York, 1965.

T. J. Dolan, *Fusion Research,* Vol. 1: *Principles;* Vol. 2: *Experiments;* Vol. 3: *Technology;* Pergamon Press, Elmsford, NY, 1982.

J. J. Duderstadt and L. J. Moses, *Inertial Confinement Fusion,* John Wiley & Sons, New York, 1982.

Samuel Glasstone and Ralph H. Loveberg, *Controlled Thermonuclear Reactions. An Introduction to Theory and Experiment,* D. Van Nostrand, Princeton, NJ, 1960.

E. H. Holt and R. E. Haskell, *Foundations of Plasma Dynamics,* Macmillan, New York, 1965.

Terry Kammash, *Fusion Reactor Physics: Principles and Technology,* Ann Arbor Science Publishers, Ann Arbor, MI, 1975.

M. A. Kettani and M. F. Hoyaux, *Plasma Engineering,* Halsted Press, New York, 1973.

E. W. Laing, *Plasma Physics,* Crane, Russak and Co., New York, 1977.

Conrad L. Longmire, *Elementary Plasma Physics,* Interscience Publishers, New York, 1963.

George H. Miley, *Direct Conversion of Nuclear Radiation Energy,* American Nuclear Society, La Grange Park, IL, 1970.

George H. Miley, *Fusion Energy Conversion,* American Nuclear Society, La Grange Park, IL, 1976.

G. C. Pomraning, *The Equations of Radiation Hydrodynamics,* Pergamon Press, Elmsford, NY, 1973.

David J. Rose and Melville Clark, Jr., *Plasmas and Controlled Fusion,* MIT Press and Wiley, New York, 1961.

I. P. Shkarovsky, T. W. Johnston, and M. P. Bachynski, *The Particle Kinetics of Plasmas,* Addison-Wesley Publishing Co., Reading, MA, 1966.

Weston M. Stacey, *Fusion Plasma Analysis,* Wiley-Interscience, New York, 1981.

T. H. Stix, *The Theory of Plasma Waves,* McGraw-Hill, New York, 1962.

B. S. Tannenbaum, *Plasma Physics,* McGraw-Hill, New York, 1967.

Derek A. Tidman and Nicholas A. Krall, *Shock Waves in Collisionless Plasmas,* Wiley-Interscience, New York, 1971.

Ya. B. Zeldovich and Yu. P. Raizer, *Physics of Shock Waves and High-Temperature Hydrodynamic Phenomena,* Academic Press, New York, 1967.

Radiation and Shielding

J. A. Ayres, ed., *Decontamination of Nuclear Reactors and Equipment,* John Wiley & Sons, New York, 1970.

John A. S. Adams and Wayne M. Lowder, eds., *The Natural Radiation Environment,*The University of Chicago Press, Chicago, 1964.

Howard L. Andrews, *Radiation Biophysics,* Prentice-Hall, Englewood Cliffs, NJ, 1961.

F. H. Attix and W. C. Roesch, eds., *Radiation Dosimetry,* Vol. I: *Fundamentals;* Vol. II: *Instrumentation;* Vol. III: *Sources, Fields, Measurements, Applications;* 2nd edit., Academic Press, New York, 1969; Suppl. 1, 1974.

M. Barbier, *Induced Radioactivity,* North-Holland Publishing Co., Amsterdam, 1969.

Glenn F. Cameron, *Radiation Detection and Measurement,* John Wiley & Sons, New York, 1979.

R. E. Chrien and W. R. Kane, eds., *Neutron Capture Gamma-Ray Spectroscopy,* Plenum Press, New York, 1979.

C. E. Crouthamel, ed., *Applied Gamma-Ray Spectrometry,* Pergamon Press, Elmsford, NY, 1960.

Geoffrey G. Eichholz, *Principles of Nuclear Radiation Detection,* Ann Arbor Science Publishers, Ann Arbor, MI, 1982.

Merril Eisenbund, *Environmental Radioactivity,* McGraw-Hill, New York, 1963.

Gerhard Erdtmann, *Neutron Activation Tables,* Verlag Chemie, Weinheim, W. Germany, 1976.

R. P. Gardner and R. L. Ely, *Radioisotope Measurement Applications in Engineering,* Reinhold Publishing Co., New York, 1967.

R. D. Gibbs and K. Way, "Directory to Nuclear Data Tabulations," U.S. Government Printing Office, Washington, DC, 1958.

R. Gibson, ed., *The Safe Transport of Radioactive Materials,* Pergamon Press, Elmsford, NY, 1966.

H. Goldstein, *Fundamental Aspects of Reactor Shielding,* Addison-Wesley Publishing Co., Reading, MA, 1959.

"Handbook on Nuclear Activation Cross Sections," International Atomic Energy Agency, Vienna, 1974.

Hugh F. Henry, *Fundamentals of Radiation Protection,* Wiley-Interscience, New York, 1969.

R. G. Jaeger, *Principles of Radiation Protection Engineering,* McGraw-Hill, New York, 1965.

R. G. Jaeger, ed., *Engineering Compendium on Radiation Shielding,* 3 vols., Springer-Verlag, New York and Berlin, 1968.

David C. Kocher, "Radioactive Decay Data Tables: A Handbook of Decay Data for Application to Radiation Dosimetry and Radiological Assessments," Technical Information Center, Oak Ridge, TN, 1981.

J. Kohl, R. D. Zentner, and H. R. Lukens, *Radioisotope Applications Engineering,* D. Van Nostrand, Princeton, NJ, 1961.

Paul Kruger, *Principles of Activation Analysis,* Wiley-Interscience, New York, 1971.

Ralph E. Lapp and Howard L. Andrews, *Nuclear Radiation Physics,* 4th edit., Prentice-Hall, Englewood Cliffs, NJ, 1972.

C. M. Lederer and V. S. Shirley, *Table of Isotopes,* 7th edit., John Wiley & Sons, New York, 1978.

Thomas D. Luckey, *Hormesis with Ionizing Radiation,* CRC Press, Boca Raton, FL, 1980.

John W. McKlveen, *Fast Neutron Activation Analysis,* Ann Arbor Science Publishers, Ann Arbor, MI, 1981.

A. Alan Moghissi, ed., *Radioactive Waste Containing Transuranic Elements,* Pergamon Press, Elmsford, NY, 1983.

K. Z. Morgan and J. E. Turner, eds., *Principles of Radiation Protection,* John Wiley & Sons, New York, 1967.

S. S. Nargolwalla and E. P. Przybylowicz, *Activation Analysis with Neutron Generators,* Wiley-Interscience, New York, 1973.

Marilyn M. Osterhout, ed., *Decontamination and Decomissioning of Nuclear Facilities,* Plenum Publishing Co., New York, 1980.

Ralph T. Overman and Herbert M. Clark, *Radioisotope Techniques,* McGraw-Hill, New York, 1960.

B. T. Price, C. C. Horton, and K. T. Spinney, *Radiation Shielding,* Pergamon Press, New York, 1957.

William J. Price, *Nuclear Radiation Detection,* 2nd edit., McGraw-Hill, New York, 1964.

A. Edward Profio, *Radiation Shielding and Dosimetry,* John Wiley & Sons, New York, 1979.

D. J. Rees, *Health Physics, Principles of Radiation Protection,* The MIT Press, Cambridge, MA, 1967.

L. W. Rickets, J. E. Bridges, and J. Miletta, *EMP Radiation and Protective Techniques,* John Wiley & Sons, New York, 1976.

T. Rockwell, ed., *Reactor Shielding Design Manual,* D. Van Nostrand, Princeton, NJ, 1956.

N. M. Schaeffer, ed., "Reactor Shielding for Nuclear Engineers," U.S. Atomic Energy Commission, 1973.

D. Shirk, F. Hsue, T. K. Li, and T. Canada, *Radioelement Analysis,* Ann Arbor Science Publishers, Ann Arbor, MI, 1980.

D. De Soete, R. Gijbels, and J. Hoste, *Neutron Activation Analysis,* Vol. 34, Chemical Analysis Series, Wiley-Interscience, New York, 1972.

L. V. Spencer, A. B. Chilton, and C. M. Eisenhauer, "Structure Shielding Against Fallout Gamma Rays from Nuclear Detonations," U.S. National Bureau of Standards, Washington, DC, 1980.

Donald C. Stewart, *Handling Radioactivity,* John Wiley & Sons, New York, 1981.

Ellery Strom and Harvey I. Israel, *Photon Cross Sections from 1 KeV to 100 MeV for Elements Z=1 to Z=100,* Nuclear Data Tables A7, Academic Press, New York, 1970.

W. H. Sullivan, *Trilinear Chart of Nuclear Species,* 2nd edit., John Wiley & Sons, New York, 1957.

J. Wood, *Computational Methods in Reactor Shielding,* Pergamon Press, Elmsford, NY, 1981.

Reactor Fuel Cycles

C. B. Amphlett, *Treatment and Disposal of Radioactive Wastes,* Pergamon Press, Elmsford, NY, 1961.

J. Belle, ed. "Uranium Dioxide: Properties and Nuclear Applications," U.S. Government Printing Office, Washington, DC, 1961.

M. Benedict and T. H. Pigford, *Nuclear Chemical Engineering,* McGraw-Hill, New York, 1957.

M. Benedict, T. H. Pigford, and H. Levi, *Nuclear Chemical Engineering,* 2nd edit., McGraw-Hill, New York, 1981.

A. G. Blasewitz, John M. Davis, and Marilynne R. Smith, *Radioactive Wastes,* Battelle Press, Columbus, OH 1982.

A. S. Coffinberry and W. N. Miner, eds., *The Metal Plutonium,* The University of Chicago Press, Chicago, 1961.

J. C. Collins, ed., *Radioactive Wastes. Their Treatment and Disposal,* Wiley, New York, 1960.

E. H. P. Cordfunke, *The Chemistry of Uranium,* American Elsevier Publishing Co., New York, 1970.

John F. Flagg, ed., *Chemical Processing of Reactor Fuels,* Academic Press, New York, 1961.

E. Glueckauf, ed., *Atomic Energy Waste, Its Nature, Use and Disposal,* Interscience, New York, 1961.

Harvey W. Graves, *Nuclear Fuel Management,* John Wiley & Sons, New York, 1979.

Earl K. Hyde, I. Pearlman, and Glen T. Seaborg, *The Nuclear Properties of the Heavy Elements,* 3 vols., Prentice-Hall, Englewood Cliffs, NJ, 1964.

Albert R. Kaufmann, ed., *Nuclear Reactor Fuel Elements, Metallurgy and Fabrication,* Wiley-Interscience, New York, 1962.

Justin T. Long, *Engineering for Nuclear Fuel Reprocessing,* American Nuclear Society, La Grange Park, IL, 1978.

James E. Lovett, *Nuclear Materials–Accountability Management Safeguards* American Nuclear Society, La Grange Park, IL, 1974.

Ralph F. Lumb, ed. *Management of Nuclear Materials,* D. Van Nostrand Company, Princeton, NJ, 1960.

Y. Marcus and A. S. Kertes, *Ion Exchange and Solvent Extraction of Metal Complexes,* Wiley-Interscience, New York, 1969.

C. A. Mawson, *Management of Radioactive Wastes,* D. Van Nostrand, Princeton, NJ, 1965.

D. R. Olander, "Fundamental Aspects of Nuclear Reactor Fuel Elements," USERDA, Oak Ridge, TN, 1976.

C. R. Russell, *Reactor Safeguards,* Pergamon Press, Elmsford, NY, 1962.

Benjamin Sanders, *Safeguards Against Nuclear Proliferation,* The MIT Press, Cambridge, MA, 1975.

Glen T. Seaborg, *Man-Made Transuranium Elements,* Prentice-Hall, Englewood Cliffs, NJ, 1963.

R. Sher and S. Untermyer, *The Detection of Fissionable Materials by Nondestructive Means,* American Nuclear Society, 1980.

P. Silvennoinen, *Reactor Core Fuel Management,* Pergamon Press, Elmsford, NY, 1976.

P. Silvennoinen, *Nuclear Fuel Cycle Optimization,* Pergamon Press, Elmsford, NY, 1982.

Herbert Sorantin, *Determination of Uranium and Plutonium in Nuclear Fuels,* Verlag Chemie, Weinheim, W. Germany, 1975.

Conrad C. Straub, "Low Level Radioactive Wastes; Their Handling, Treatment and Disposal," Bureau of Technical Information, U.S. Atomic Energy Commission, Washington, DC, 1964.

M. Taube, *Plutonium,* Macmillan, New York, 1964.

L. S. Tong, *Engineering for Nuclear Fuel Reprocessing,* Gordon & Breach, Science Publishers, New York, 1967.

Stelio Villani, *Isotope Separation,* American Nuclear Society, La Grange Park, IL, 1976.

S. Villiani, ed., *Uranium Enrichment,* Springer-Verlag, New York, 1979.

L. R. Weissert and G. Schileo, *Fabrication of Thorium Fuel Elements,* American Nuclear Society, 1968.

O. J. Wick, ed., *Plutonium Handbook,* Vol. 1, Gordon & Breach, Science Publishers, New York, 1967.

W. D. Wilkinson, *Uranium Metallurgy,* Vol. I: *Uranium Process Metallurgy;* Vol. II: *Uranium Corrosion and Alloys;* Wiley-Interscience, New York, 1962.

A. D. Wordsworth, *Nuclear Fuel Handling,* Butterworth Publishers, London, 1963.

Raymond G. Wymer, ed., *Light Water Reactor Nuclear Fuel Cycle,* CRC Press, Boca Raton, FL, 1981.

General Topics, Policy

E. William Colglazier, *Politics of Nuclear Waste,* Pergamon Press, Elmsford, NY, 1982.

Charles Fairhurst and Donald Gills, *Managing Nuclear Waste,* Pergamon Press, Elmsford, NY, 1980.

Samuel Glasstone, *Sourcebook on Atomic Energy,* 3rd edit., Van Nostrand, Princeton, NJ, 1967.

Bertrand Goldschmidt, *The Atomic Complex,* American Nuclear Society, LaGrange Park, IL, 1982.

Peter Goodchild, *J. Robert Oppenheimer, Shatterer of Worlds,* Houghton Mifflin Co., Boston, MA, 1981.

John F. Hogerton, *The Atomic Energy Deskbook,* Reinhold Publishing Co., New York, 1963.

Lee M. Hydeman and William H. Berman, "International Control of Nuclear Maritime Activities," The University of Michigan Law School, Ann Arbor, MI, 1960.

Samuel McCracken, *The War Against the Atom,* Basic Books, New York, 1982.

L. M. Muntzing, ed., *Nuclear Power and its Regulation in the United States,* Pergamon Press, Elmsford, NY, 1981.

Eric S. Pedersen, *Nuclear Power,* Vol. 1: *Nuclear Power Plant Design;* Vol. 2: *Nuclear Power Project Management;* Ann Arbor Science Publishers, Ann Arbor, MI, 1978–1982.

Stanley Reigel and P. John Owen, *Administrative Law: The Law of Government Agencies,* Ann Arbor Science Publishers, Ann Arbor, MI, 1982.

E. Segre, ed. *The Collected Papers of Enrico Fermi, Volume 1, 1921–1938*, The University of Chicago Press, Chicago, 1962.

E. N. Shaw, *High Temperature and Cold Comfort from Nuclear Politics, The Official History of the DRAGON Project,* Pergamon Press, Elmsford, NY, 1982.

Hugh B. Stewart, *Transitional Energy Policy 1980–2030, Alternative Nuclear Technologies,* Pergamon Press, New York, 1981.

Alexander De Volpi, *Proliferation, Plutonium and Policy,* Pergamon Press, New York, 1979,

Jerry L. Weinstein, ed., *Progress in Nuclear Energy. Law and Administration,* Vol. 4, Pergamon Press, Elmsford, NY, 1966.

Charles E. Whittle et al., *Economic and Environmental Impacts of a U.S. Nuclear Moratorium, 1985–2010,* The MIT Press, Cambridge, MA, 1979.

F. C. Williams, *Nuclear Nonproliferation, The Spent Fuel Problem,* Pergamon Press, Elmsford, NY, 1979.

Index